Rocks & Minerals of Western North America

Azurite coating Malachite (Bisbee, Arizona), Mineral Museum, Butte, MT.

J. Duane Sept

Calypso Publishing

Text copyright © 2013 J. Duane Sept
Photographs & Illustrations copyright © 2013 J. Duane Sept
Third Printing 2017

All rights reserved. No part of this publication may be reproduced, stored in a retrieval system or transmitted, in any form or by any means, without prior permission of the publisher, except by a reviewer who may quote brief passages in a review.

Front cover photos:
Actinolite Nephrite Polar Mine, near Dease Lake, British Columbia
Rhodonite with Quartz, Salt Spring Island, British Columbia
Quartz, hollow Amethyst egg within an egg, Dugway, Willow Springs, Utah
Topaz, San Diego County, California

Back cover photo:
Peach Quartz, Chalcedony lining, Keokuk Geode, Princeton, British Columbia

Printed & bound in China.

Calypso Publishing
P. O. Box 1141
Sechelt, BC, Canada
V0N 3A0
www.calypso-publishing.com

Duane Sept Photography
www.septphoto.com

Library and Archives Canada Cataloguing in Publication

Sept, J. Duane, 1950-
Rocks & minerals of western North America / J. Duane Sept.

Includes bibliographical references and index.
ISBN 978-0-9739819-7-1

1. Rocks--Canada, Western--Identification. 2. Rocks--West (U.S.)--Identification. 3. Minerals--Canada, Western--Identification. 4. Minerals--West (U.S.)--Identification. I. Title.

QE71.S46 2013 552.097 C2013-900135-2

Contents

Introduction

The world of rocks and minerals is an amazing one to discover. From the spectacular landscapes we observe to the tiny crystals we unearth, rocks and minerals are all around us, full of beauty, surprises and stories of the planet Earth.

Approximately 2,500 different minerals have been described in North America, and new ones are being discovered each year. This introductory guide focuses on some of the common rocks and minerals that can be seen in western North America.

One of the first things to notice about rocks and minerals is that some feel extremely heavy for their size (e.g., galena), while others feel extremely light (e.g., coal). Galena is heavy because it has small, dense molecules that are packed together tightly, whereas coal is light because it has large molecules with lots of air. Most rocks and minerals are somewhere in between.

The Difference Between Rocks and Minerals

A mineral is a homogenous, naturally occurring inorganic solid with a defined chemical composition.

A rock is almost always a combination of a few essential minerals combined with additional minerals. A few rocks, such as limestone, are made almost entirely of one mineral.

A gem, or gemstone, is a mineral or rock valued for its beauty, rarity and hardness. Many, but not all, gemstones are crystals. Gemstones are used in making fashion clothing and jewelry.

Precious and semi-precious gemstones are terms that have been used to rank gemstones as to quality and rarity. These designations change over time, however, with shifts in rarity and popularity. The quality of individual specimens can also make a difference—often changing a precious stone to a semi-precious one. Therefore, these terms have little meaning today.

Types of Minerals

Minerals are divided into several major groups. See individual sections for details on each group.

Sulfides (p. 23), Sulfosalts (p. 29), Oxides (p. 31), Carbonates (p. 32), Sulfates (p. 37), Silicates (p. 42)

Types of Rocks

Rocks are classified into three major groups: igneous, sedimentary and metamorphic.

Igneous Rocks

Igneous rocks are volcanic in nature, forming from magma (molten rock) as it cools after eruption. All rocks that are transparent or glass-like are volcanic in nature. There are two groups of igneous rocks: extrusive and intrusive.

Extrusive igneous rocks.
When lavas flow out onto the surface of the Earth. and cool, they become extrusive igneous rocks, or extrusives. Cooling on the surface occurs very quickly, and the crystals produced are very small because they do not have enough time to grow large. A good example is basalt.

Intrusive (plutonic) igneous rocks.
When lavas flow deep into fissures, they cool slowly and the crystals grow much larger, forming intrusive igneous rocks, also known as plutonics or intrusives. Their composition depends upon the chemical makeup of the molten rock, which in turn depends largely upon the elements present in the magma at the time of eruption.

Sedimentary Rocks

Sedimentary rocks are formed by the erosion, transportation and deposition of existing rocks. They tend to be easy to recognize because they normally have a layered structure. There are three groups of sedimentary rocks.

Detrital (clastic) rock.
Rock formed by the accumulation of fragments derived from pre-existing rocks and minerals, which are deposited by mechanical agents such as water, wind and ice. Examples include sandstone and conglomerate.

Chemical rock.
When minerals are precipitated directly from a saline or freshwater solution, the chemical sediments form rock. Examples include rock salt and gypsum rock.

Biogenic (organic) rock.
Organic sedimentary rock formed directly or indirectly through the action of plants or animals. Examples include coal and coral limestone.

Metamorphic Rocks

Metamorphic rocks are created by the physical and chemical changes of existing igneous and sedimentary rocks, caused mainly by the forces of heat and pressure. There are three groups of metamorphic rocks.

Contact metamorphic rock.
Contact metamorphism is caused by direct heat. which usually creates crystalline rock. Examples include hornfels.

Regional metamorphic rock.
Regional metamorphism is caused by heat and pressure, which produces foliation (parallel layers) or cleavage in rocks where the minerals have been aligned (layered).

Cataclastic metamorphic rock.
Cataclastic metamorphism takes place where rocks are being faulted and sheared. This is not a common type of rock.

How To Use This Book

This guide describes various properties of each mineral and rock, to help you in identifying them.

Minerals

Mineral Name

The common name that is accepted for this mineral. If there is more than one common name, the favored one will be shown here, and the other will be mentioned.

Composition

The chemical composition and its scientific name.

Classification

The group of the mineral.

Color

The color(s) of the mineral.

Luster

The manner in which a mineral shines, affected by the way that light is reflected from the surface. For instance, the mineral may be:

- vitreous (glassy)
- silky (finely fibrous, in parallel bands of needles)
- pearly (resembling the flat surfaces of an iridescent shell).

Streak

The color of the mineral's powder. To obtain a sample, rub the specimen across the surface of an unglazed porcelain tile.

Fracture

The irregularly shaped pieces of broken crystals.

Crystals

Description and illustration of the crystal structure of the mineral. A crystal is a solid material whose molecules are in one of several ordered patterns (below). The various crystal habits (predominant forms) are also included in this section.

Isometric (Cubic)
A structure characterized by three equal axes. It makes a cubic shape, with all sides having the same dimensions. The possible forms produced include cubes, octahedrons and dodecahedrons.

Tetragonal
A structure characterized by two equal axes and a third axis having a different length. All angles are 90 degrees, so this structure makes an elongated cubic shape with the end view being square and the side views being rectangular. The possible forms produced include dipyramid, prism and basal pinacoid.

Hexagonal
A structure characterized by three equal axes all crossing at 60 degrees to each other, and another axis having a different length and crossing at 90 degrees, making an elongated or shortened hexagonal form with flattened ends. The possible forms produced include three- or six-sided prisms, pyramids and rhombohedrons.

Trigonal
A structure characterized by three equal axes—a, b and c—and the e axis having a different length. All angles are at 60 degrees, so this structure makes an elongated or shortened hexagonal shape from any side, and a perfect hexagonal shape from the end. The possible forms produced include four-sided prisms, pyramids and dipyramids.

Orthorhombic
A structure characterized by no equal axes—all three are a different length. All angles are at 90 degrees, so this structure makes a rectangular shape from any side. The possible forms produced include four-sided prisms, pyramids and pinacoids.

Monoclinic
A structure characterized by no equal axes—all three are different lengths. One angle is at 90 degrees, so this structure makes a rectangular shape with a wedge-shaped end. The possible forms produced include prisms and pinacoids.

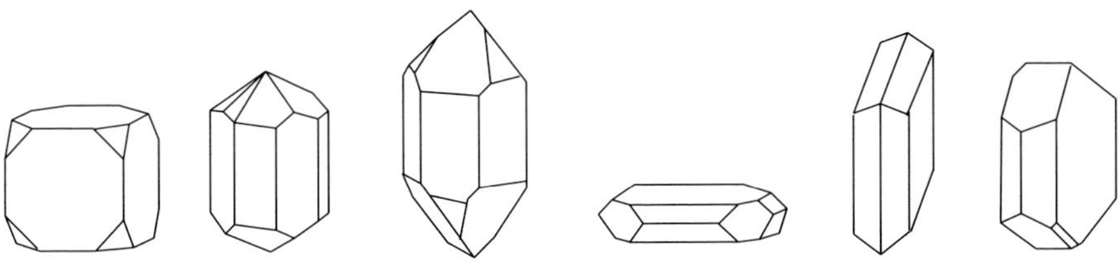

Isometric *Tetragonal* *Hexagonal/Trigonal* *Orthorhombic* *Monoclinic* *Triclinic*

Triclinic

A structure characterized by no equal axes—all three are different lengths. There is no 90-degree angle, so this structure makes an irregular shape with no square corners. The possible forms produced are pinacoids.

Formation

A description of how each mineral was formed.

Group

The group and type of mineral.

Hardness

The property of resistance to abrasion or scratching, measured by the Mohs table of hardness (below). Because rocks are made up of more than one mineral, this measurement is not valid for rocks.

Hardness Measurement

Mohs Table of Hardness	Scratched with
A little over 2	Fingernail
Approximately 3	Copper coin
A little over 5	Pocket knife
5.5	Window glass
6.5	Steel file

Mohs Table of Hardness

1. TALC
2. GYPSUM
3. CALCITE
4. FLUORITE
5. APATITE
6. ORTHOCLASE
7. QUARTZ
8. TOPAZ
9. CORUNDUM
10. DIAMOND

Cleavage

Crystal fragments with essentially flat surfaces parallel to crystallographic planes (faces).

Notes

Interesting observations and information on the mineral.

Rocks

Classification

The group and subgroup of the rock.

Color

Color or possible colors of the rock.

Texture

Surface appearance of a homogenous rock sample.

Origin

How this rock was formed.

Gain Size
A visual description of the grain of the rock. Some are quite coarse while others have virtually no visible grain.
Notes
Interesting observations and information on the rock.

Collecting Rocks and Minerals

Where to Collect Specimens

There are many good places to look for both rocks and minerals, including pebble beaches—an excellent location to search for specimens—and lakesides, riverbanks and the bases of cliffs. When pebbles are wet, their colors are more distinct, so look for pebbles wet with rain or near water to help in separating the interesting specimens from the not-so-interesting ones.

Old quarries, mine workings and mine dumps are usually productive sites to collect minerals, but be very careful! Shafts and tunnels in disused mines are not always obvious and can be dangerous.

Collecting Equipment

You need only a few inexpensive tools to collect rocks and minerals: a geological hammer, a pocket knife, cold chisels, a magnifying glass or hand lens (8x to 10x), a carrying bag, a notepad and a pencil.

Protective clothing is also essential. If you are working below a high cliff or quarry face, a hard hat is necessary. Safety goggles will shield your eyes from chips of rock flying off specimens when you use your hammer. Gloves will protect your hands.

Care of the Collection

Once you have spent the time and effort to locate and collect your specimens, take good care of your collection to make it worth keeping. Carefully wrap and label each specimen collected in the field. Do so immediately, rather than waiting until you get home, if you visit several locations on a single trip.

Clean your specimens, and make a permanent label for each one. Organize the minerals and rocks for storage and possible display. As you collect each specimen, record the location and date on a card, and create a card file or a computer database. Keep each specimen in an individual tray or box except when they are on display.

Lapidary Arts

Many collectors of rocks and minerals find that their interest and effort is rewarded through the creation of beautiful things produced in the lapidary arts (lapidary means "concerned with stones"). These arts include tumbling, and the cutting and shaping of cabochons and cameos. As with many other pastimes, the best way to learn more about the techniques and traditions of these activities is to join a local lapidary or rock and mineral club. Some organizations have lapidary equipment available to members. An amazing number of clubs are active throughout the west. See page 88 for websites of clubs in your region.

Yellow Feather Jasper (UT)

Serpentine (BC)

Rhodonite (BC)

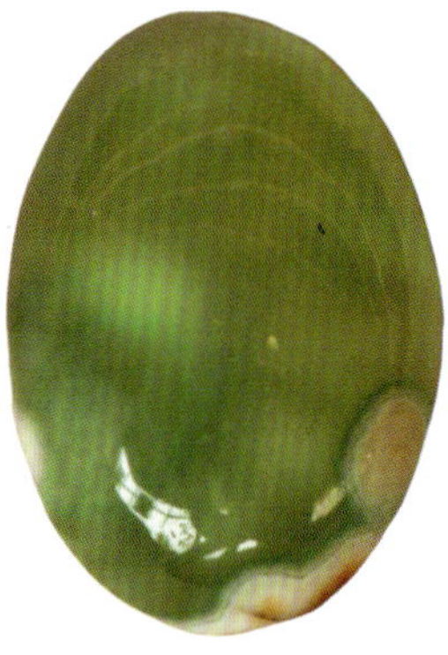

Blue Opal & Agate (BC)

Quick Photo Guide to Minerals

Gold p. 21

Silver p. 22

Copper p. 22

Galena p. 23

Cinnabar p. 24

Sphalerite p. 24

Bornite p. 25

Chalcopyrite p. 25

Covellite p. 26

Orpiment p. 26

Molybdenite p. 27

Pyrite p. 27

Quick Photo Guide to Minerals

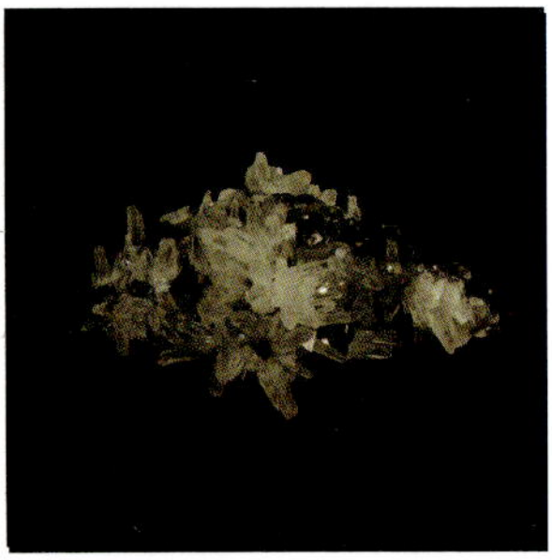
Arsenopyrite p. 28

Realgar p. 28

Stibnite p. 29

Enargite p. 29

Tetrahedrite p. 30

Tennantite p. 30

Cuprite p. 31

Magnetite p. 32

Calcite p. 32

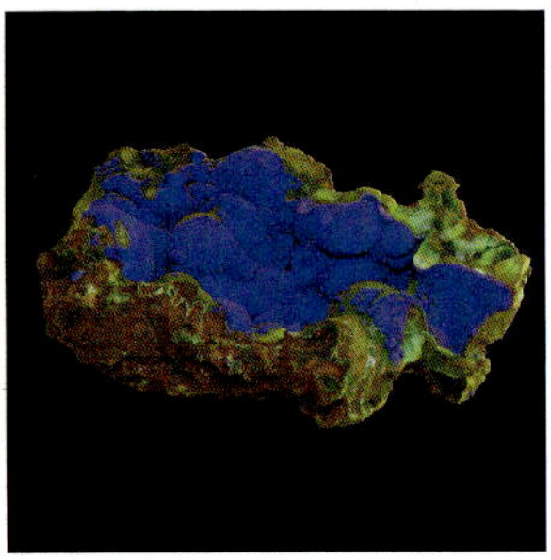
Azurite p. 33

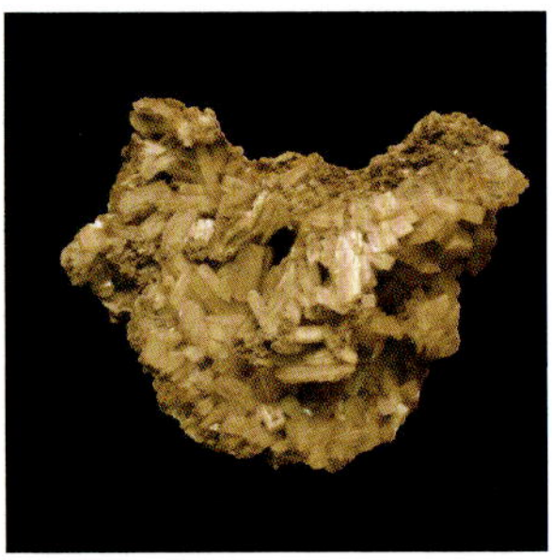
Cerussite p. 33

Dolomite (Mineral) p. 34

Quick Photo Guide to Minerals

Magnesite p. 34

Malachite p. 35

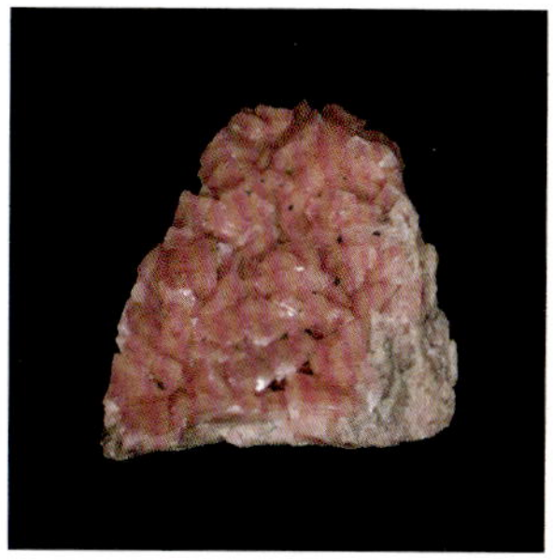

Rhodochrosite p. 35

Borax p. 36

Barite p. 37

Chalcanthite p. 37

Cyanotrichite p. 38

Linarite p. 38

Scheelite p. 39

Wulfenite p. 39

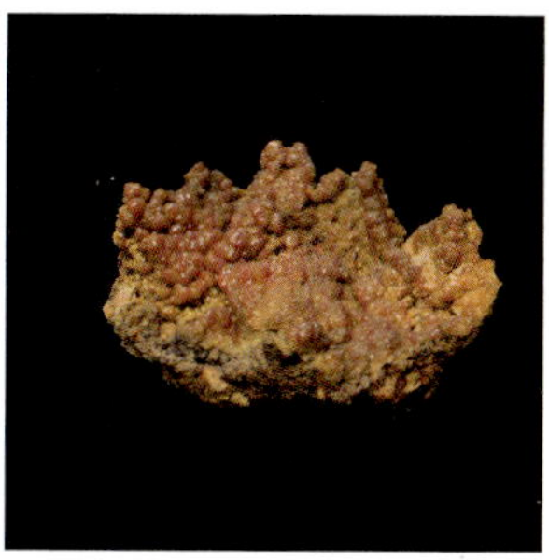

Pyromorphite p. 40

Variscite p. 40

Quick Photo Guide to Minerals

Vivianite p. 41

Vanadinite p. 41

Actinolite Nephrite p. 42

Chalcedony p. 44

Almandine (Garnet) p. 46

Anorthite p. 46

Anthophyllite p. 47

Beryl p. 47

Chabazite p. 48

Chrysotile p. 48

Clinochlore p. 49

Elbaite Toumaline p. 49

Quick Photo Guide to Minerals

Epidote p. 50

Hemimorphite p. 50

Heulandite p. 51

Ilvaite p. 51

Kyanite p. 52

Laumontite p. 52

Lepidolite p. 53

Muscovite p. 53

Natrolite p. 54

Neptunite p. 54

Orthoclase Feldspar p. 55

Phlogopite p. 55

Quick Photo Guide to Minerals

Quartz p. 56

Rhodonite p. 58

Serpentine p. 58

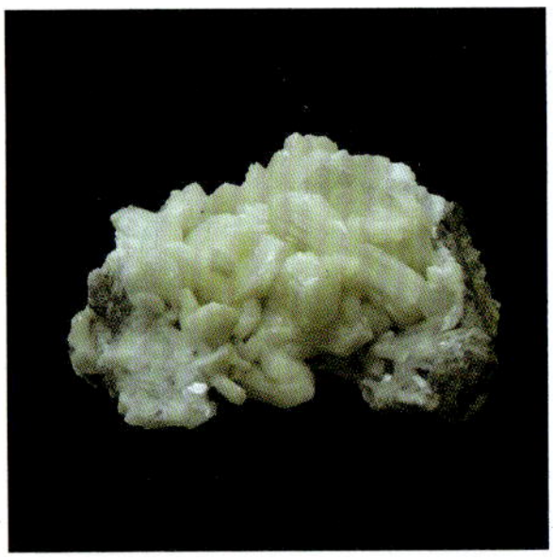

Stilbite p. 59

Talc p. 59

Thomsonite p. 60

Titanite p. 61

Topaz p. 61

Tremolite p. 62

Zoisite p. 62

Amber p. 63

Opal p. 64

Quick Photo Guide to Rocks

Andesite p. 67

Anorthosite p. 68

Diorite p. 68

Basalt p. 69

White Granite p. 70

Pink Granite p. 70

Granodiorite p. 71

Gabbro p. 71

Obsidian p. 72

Pegmatite p. 72

Porphyry p. 73

Pumice p. 74

Quick Photo Guide to Rocks

Rhyolite p. 74

Arkose p. 75

Bituminous Coal p. 76

Breccia p. 76

Gypsum Rock p. 77

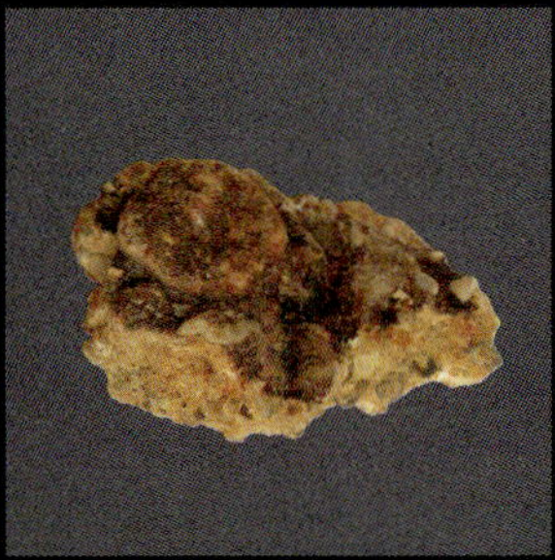
Conglomerate p. 77

Dolomite Rock p. 78

Limestone p. 78

Fossiliferous Limestone p.79

Shale p. 79

Sandstone p. 80

Fossiliferous Sandstone p.80

Quick Photo Guide to Rocks

Siltstone p. 81

Mudstone p. 81

Amphibolite p. 82

Eclogite p. 83

Gneiss p. 83

Hornfels p. 84

Phyllite p. 84

Marble p. 85

Quartzite p. 86

Schist p. 86

Serpentinite p. 87

Slate p. 87

MINERALS

Barite and Calcite, Stalactitic Calcite, Montana.

NATIVE ELEMENTS: METALS

Most minerals are found as compounds, in combinations of chemicals. Some "native elements," however, occur by themselves in relatively pure form. There are three major groups: the metals, the semimetals and the nonmetals.
Metals are more numerous, and are readily recognized because they have a metallic appearance. The metals are most easily identified by their color and malleability (easily deformed by pounding, without crumbling).

Gold

The Highland Centennial Gold Nugget.

Composition Au • Gold **Classification:** Minerals – Native Elements. **Color** Golden yellow. **Luster** Metallic. **Transparency** Opaque. **Streak** Golden yellow. **Fracture** Hackly (rough surface with small lumps). **Crystals** Rare and tiny, cubic or octahedral; normally forms grains and nuggets or clusters. **Formation** Forms in hydrothermal veins, often with quartz and sulfides; also in placer deposits of sand, and in sandstone and conglomerate. **Hardness** 2.5-3. **Cleavage** None.
Notes Gold has increased in value tremendously over the years and is recognized as a relatively safe commodity to invest in. Gold is instantly recognizable. A few other minerals, including pyrites, other sulphides and some tellurides, can be mistaken for gold; but only gold forms gold grains and nuggets.
Gold mines in western North America include sites in the Yukon, California, Colorado and South Dakota. Since gold is now mined with large machinery, the chance of discovering large crystallized gold specimens is greatly reduced.

The "Fricot Nugget" (Crystallized Gold, 201 Troy Ounces).

Silver

Composition Ag • Silver
Classification Minerals – Native Elements. **Color** Silver-white. **Luster** Metallic. **Transparency** Opaque. **Streak** Silvery white. **Fracture** Hackly (jagged). **Crystals** Isometric; crystals (rare) are cubes. **Formation** Forms in volcanic rocks and hydrothermal veins. **Hardness** 2.5-3. **Cleavage** None.
Notes Silver deposits are often found with gold or metallic sulfides. When found with gold, silver forms 20 to 25% of the gold and silver alloy, called electrum.

Silver tarnishes on exposure to the atmosphere, as is often noticed when the good table silverware is brought out for special occasions. Silver is also an excellent conductor of electricity and heat.

Bunker Hill Mine, Kellogg, Idaho.

Copper

Yellowknife, NWT.

Composition Cu • Copper
Classification: Minerals – Native Elements. **Color** Reddish gold. **Luster** Metallic. **Transparency** Opaque. **Streak** Reddish gold. **Fracture** Hackly (jagged). **Crystals** Typically in distorted, complex crystals, with cubes, dodecahedrons and octahedrons. Often in hackly masses. **Formation** Forms in volcanic rocks (basic), and in sulfide deposits. **Hardness** 2.5-3. **Cleavage** None.
Notes Copper is a very soft metal and is only occasionally found in the pure native form. The first knives were made by humans approximately 6,000 years ago, from this pure form of copper. Copper is rare as a native element, and chalcopyrite and similar ores provide most of the copper used today.

British Columbia is one of the largest producers of copper in the world. In western North America, native copper has also been discovered in Oregon and Alaska.

SULFIDES

Sulfides (also "sulphides") are compounds of sulfur plus one or two metallic or semimetallic elements. As a result, a sulfide usually has a metallic luster. Well-formed, very symmetrical crystals are the rule for this group.

Galena

Composition PbS • Lead sulfide **Classification** Minerals – Sulfides. **Color** Dark lead-gray. **Luster** Bright metallic. **Transparency** Opaque. **Streak** Lead gray. **Fracture** Even (rarely observed). **Crystals** Isometric; normally in cubes or combinations of cubes and octahedrons. **Formation** Forms in hydrothermal veins when hot fluids find their way to the Earth's crust. **Hardness** 2.5. **Cleavage** Perfect. **Notes** Galena is a noticeably heavy ore as it is made up of 86.6% lead. It often contains several impurities, including silver, arsenic, antimony and others. Because large quantities of galena are mined annually, it is a major source of silver.

Galena on Calcite.

Silver Hill Mine, Tulameen, British Columbia.

Historically, galena is the main ore that was mined for lead. it has been mined extensively in British Columbia. In years past, galena was well known for its use as the crystal for the radio.

Cinnabar

New Almaden mine, Santa Clara County, California.

Composition HgS • Mercury sulfide **Classification** Minerals – Sulfides. **Color** Scarlet to brick red. **Luster** Adamantine (bright mineral luster), submetallic or dull. **Transparency** Transparent to opaque. **Streak** Scarlet. **Fracture** Uneven to splintery. **Crystals** Hexagonal; rhombohedral, and prismatic or thick tabular. **Formation** Forms with realgar and pyrite around hot springs and volcanic vents. **Hardness** 2-2.5. **Cleavage** Perfect.

Notes Cinnabar is the main source of mercury, which is rarely found in nature. However, mercury is liquid at room temperature; therefore, mercury is only found in tiny blobs in crevices of cinnabar. In North America, important cinnabar deposits occur in California, Nevada and Oregon.

Caution Carefully wash your hands after handling cinnabar. Mercury is poisonous.

Sphalerite

Composition (Zn,Fe)S • Zinc iron sulfide

Classification Minerals – Sulfides. Color Yellow, brown, red, green, black; rarely white or gray. **Luster** Resinous. **Transparency** Translucent to transparent. **Streak** Light brown to colorless. **Fracture** Conchoidal. **Crystals** Isometric; tetrahedral and dodecahedral forms are common; faces are often rounded. **Formation** Forms in hydrothermal veins with other minerals, including dolomite, quartz, pyrite, galena, fluorite, barite and calcite. **Hardness** 3.5–4. **Cleavage** Perfect.

Notes Sphalerite is the primary ore of zinc. Under ultraviolet light, sphalerite sometimes fluoresces orange light. Triboluminescence is another phenomenon exhibited by fluorescent sphalerite—if it is lightly stroked with a knife, a stone or other hard substance, it emits flashes of orange light.

Triboluminescence is the light triggered by mechanical or electrical energy from a mechanical action such as friction.

Sphalerite & Quartz, Anselmo Mine, Butte, Montana.

Butte, Montana.

Bornite

Composition Cu_2FeS_4 • Copper iron sulphide **Classification** Minerals – Sulfides. **Color** Reddish bronze with iridescent peacock tarnish. **Luster** Metallic. **Transparency** Opaque. **Streak** Grey-black. **Fracture** Uneven to conchoidal. **Crystals** Isometric; crystals cubic (rare), dodecahedral; compact, granular or massive habits. **Formation** Forms in hydrothermal veins and in some igneous rocks. **Hardness** 3. **Cleavage** Very poor.

Notes The surface of bornite tarnishes in mere hours, leaving it coated with a dark, purplish iridescent tarnish. Bornite is an important ore for copper. Massive bornite deposits have been located at the Evergreen Mine, near Apex, Gilpin Co., Colorado; at Kennecott in the Copper River district, Alaska; at Butte, Montana; and at Texada Island, British Columbia.

Peacock ore sold to collectors as bornite is sometimes chalcopyrite (see below) whose iridescence has been enhanced with a treatment of acid. To determine whether you have bornite, scratch away the tarnish. If it is bornite, the true pink copper color will be revealed.

Chalcopyrite

Courtenay, British Columbia.

Composition $CuFeS_2$ • Copper iron sulphide **Classification** Minerals – Sulfides. **Color** Brass-yellow, tarnishes to iridescent blues, greens and purples. **Luster** Metallic. **Transparency** Opaque. **Streak** Dark green to black. **Fracture** Conchoidal, brittle. **Crystals** Tetragonal; often tetrahedral; crystal faces usually uneven; often massive, compact, granular. **Formation** Forms in a wide variety of environments, including hydrothermal veins. **Hardness** 3.5–4. **Cleavage** Poor.

Notes Chalcopyrite is one of the most important and widely distributed copper ores. Small quantities of silver and gold are often present with this ore. Its name originates from the Greek Chalkos, meaning "copper," and pyrites, which means "fiery." Chalcopyrite looks very similar to pyrite, but is more yellowish. Chalcopyrite also looks like gold, but is brittle and harder.

Covellite

Leonard Mine, Butte, Montana.

Composition CuS • Copper sulfide **Classification** Minerals – Sulfides. **Color** Iridescent indigo blue, tarnishing black or purple. **Luster** Metallic. **Transparency** Opaque. **Streak** Grey to black. **Fracture** Flaky. **Crystals** Hexagonal; plate-like; also occurs as grains and strip-like masses. **Formation** Forms in copper veins that have been altered by fluids seeping through the vein. **Hardness** 1.5–2. **Cleavage** Perfect.

Notes Covellite—or indigo copper, as it is sometimes called—typically forms in thin, hexagonal crystal plates. Some of the best examples of this uncommon mineral were found at Butte, Montana. Still better masses were discovered at Kennicott, Alaska. These two sites have produced significant amounts of covellite. Additional western North American mine sites include Summitville, Rio Grande Co., Colorado; and La Sal district, San Juan Co., Utah.

Orpiment

Treaty Creek, Stewart, British Columbia.

Composition As_2S_3 • Arsenic sulfide **Classification** Minerals – Sulfides. **Color** Orange-yellow to yellow. **Luster** Resinous to pearly. **Transparency** Transparent to translucent. **Streak** Yellow. **Fracture** Flaky. **Crystals** Monoclinic; prismatic (rare); more frequently in thin, foliated masses, or in massive or columnar habits. **Formation** Occurs in low-temperature hydrothermal veins and in the crusts deposited around hot springs. It is always associated with realgar in low-temperature veins. Realgar (AsS) (see p. 28) is a very closely related mineral. **Hardness** 1.5–2. **Cleavage** Perfect.

Notes Orpiment is a transparent to translucent mineral that contains arsenic. This fact was not considered when orpiment was used in the manufacture of women's cosmetics years ago.

Some time ago, excellent large crystals of orpiment were discovered at Mercur, Utah. When orpiment crystals are placed in museum collections they develop a white film that dulls their color, likely due to exposure to light.

Caution Wash your hands thoroughly after handling orpiment. Arsenic is poisonous.

Molybdenite

Crown Point mine, Washington.

Composition MoS_2 • Molybdenum sulphide **Classification** Minerals – Sulfides. **Color** Bluish lead grey. **Luster** Metallic. **Transparency** Opaque. **Streak** Greenish grey. **Fracture** Flaky. **Crystals** Hexagonal; crystals (rare) thin hexagonal tablets; soft sheet-like masses and crusts. **Formation** Originates in igneous rocks and pegmatites. **Hardness** 1–1.5. **Cleavage** Perfect. **Notes** Molybdenite is an extremely soft mineral that leaves dark marks, but not as dark as graphite. It has a greasy feel, similar to graphite. Noteworthy deposits in western North America have been found at Climax, Colorado; and Lake Chelan, Washington.

Pyrite

Butte, Montana.

Composition FeS_2 • Iron sulphide **Classification** Minerals – Sulfides. **Color** Pale yellow to brass-yellow. **Luster** Metallic. **Transparency** Opaque. **Streak** Greenish black. **Fracture** Conchoidal. **Crystals** Isometric; a wide variety of forms; twinning is common. **Formation** Occurs in igneous, sedimentary and metamorphic rocks. **Hardness** 6–6.5. **Cleavage** None. **Notes** Pyrite is widely known as "fool's gold" and indeed, it has even fooled many prospectors into thinking they have struck it rich. Its crystals can be truly amazing in their perfect shapes. Pyrite is one of the most common of all minerals found on Earth. To determine whether a specimen is pyrite, simply strike it with a hard metal object. If it does not give off sparks, you may have struck gold!

Pyrite with Quartz and Chalcocite, Butte, Montana.

Arsenopyrite

Composition FeAsS • Iron arsenide sulphide
Classification Minerals – Sulfides. **Color** Brassy white to gray, oxidizes to brown or pink. **Luster** Metallic. **Transparency** Opaque. **Streak** Dark grey to black. **Fracture** Uneven. **Crystals** Monoclinic; prismatic; striated faces; also granular, compact, columnar. **Formation** Occurs in hydrothermal veins, in metamorphic rocks and in basic igneous rocks. **Hardness** 5.5–6. **Cleavage** Distinct.

Notes Arsenopyrite is an ore containing arsenic, but it is rarely mined just for its arsenic. More often, it is an unwelcome byproduct in mining other metals. In order to mine the desirable metals in the ore, special processes have to be undertaken to remove this poisonous byproduct.

Caution Handle arsenopyrite with extreme caution and wash your hands thoroughly. It contains arsenic, and arsenic is poisonous.

Bluebell Mine, Riondel, British Columbia.

Realgar

Composition AsS • Arsenic sulphide
Classification Minerals – Sulfides. Color Deep red to orange, changing to yellow upon exposure to light. **Luster** Resinous, adamantine to submetallic. **Transparency** Translucent to transparent. **Streak** Orange to orange-yellow. **Fracture** Subconchoidal. **Crystals** Monoclinic; normally prismatic (crystals are rare, however); also granular and incrusting. **Formation** Forms in hydrothermal veins and around hot springs; also found with minerals of lead, silver and antimony. **Hardness** 1.5–2. **Cleavage** Good.

Getchell Mine, Humbolt Co., Nevada.

Notes Realgar is found in many locations around the world. Its bright colors change when exposed to light. In ancient times, the Chinese used realgar for carvings, which have since deteriorated and are now crumbling. Artists also used realgar as a red pigment, and today many of these old paintings have decayed to yellow or orange. Mineral collectors are well advised to keep realgar specimens in complete darkness.

Noteworthy locations for realgar in western North America include Mercur, Utah; Getchell, Nevada; and King County, Washington. Realgar is an ore of arsenic, and like other arsenic minerals, it gives off a strong smell of garlic when heated. **Caution** Wash your hands very carefully after handling realgar. It contains arsenic, and arsenic is poisonous.

Stibnite

Stibnite with Calcite, Dolomite & Quartz, Oregon.

Composition Sb_2S_3 • Antimony sulphide **Classification** Minerals – Sulfides. **Color** Steel gray to silver. **Luster** Metallic. **Transparency** Opaque. **Streak** Dark gray. **Fracture** Uneven. **Crystals** Orthorhombic; normally in well-formed crystals; variable—very large and solid to slender and fragile. **Formation** Is deposited from hot mineral springs, or forms in low-temperature hydrothermal veins. **Hardness** 2. **Cleavage** Perfect, one direction lengthwise.

Notes Stibnite is the most important ore of antimony, a silvery gray semimetal that rarely occurs as a native element.

Noteworthy long, thin, bladed or needle-like crystals are often formed. The planes inside the crystal structure slide over each other, which allows the crystals to move and bend without fracturing into long, thin crystals.

Caution Wash your hands carefully after handling stibnite. Antimony is poisonous.

SULFOSALTS

Sulfosalts (also "sulphosalts") are compounds in which one or more metallic elements combine with sulfur plus a semimetallic element (such as antimony or arsenic). Sulfosalt minerals are soft, heavy and brittle, and generally occur in small crystals or masses.

Enargite

Leonard Mine, Butte, Montana.

Composition CU_3AsS_4 • Copper arsenic sulfide **Classification** Minerals – Sulfosalts. **Color** Grayish black to iron-black. **Luster** Metallic. **Transparency** Opaque. **Streak** Dark gray to black. **Fracture** Uneven. **Crystals** Orthorhombic; tabular, prismatic; crystal faces show vertical striations; also massive or granular habits. **Formation** Occurs in hydrothermal veins, or replacement deposits. **Classification** Minerals – Sulfosalts. **Hardness** 3. **Cleavage** Perfect.

Notes This soft, heavy, fragile mineral is an important ore of copper and arsenic. Its name is derived from the Greek enargos, meaning "visible," a reference to its distinct cleavage. Western North America has many excellent spots for collecting. Enargite is found with bornite, covellite and pyrite at Butte, Montana; and accompanied by chalcocite and bornite at the Kennecott mine in the Copper River district, Alaska.

Tetrahedrite

Tetrahedrite on Quartz, Black Pine Mine, Montana.

Composition $Cu_{12}Sb_4C_{13}$ • Copper antimony sulphide **Classification** Minerals – Sulfosalts. **Color** Steel-gray to iron-black. **Luster** Metallic. **Transparency** Opaque. **Streak** Varies from black or brown to red. **Fracture** Uneven to subconchoidal. **Crystals** Isometric; tetrahedral; often twinned. **Formation** Forms in hydrothermal veins with sulfides, carbonates and others. **Hardness** 3–4. **Cleavage** Absent.

Notes Tetrahedrite is named for its sharp, distinct, well-formed tetrahedrons—its predominant crystal form. The best way to identify this mineral is by the presence of its crystals. It is also an important copper-ore mineral. Tetrahedrite has been found in the Lillooet district, British Columbia.

Tennantite

Composition $(Cu,Fe)_{12}As_4S_{13}$ • Copper iron arsenic sulfide **Classification** Minerals – Sulfosalts. **Color** Dark gray to black. **Luster** Metallic, sometimes very bright. **Transparency** Opaque. **Streak** Black, brown or dark red. **Fracture** Uneven to subconchoidal. **Crystals** Isometric; tetrahedrons; occasionally as groups of parallel crystals; also compact; frequently twinned. **Formation** Forms in hydrothermal veins, with many other minerals. **Hardness** 3–4.5. **Cleavage** None.

Butte, Montana.

Notes Tennantite was named after Smithson Tennant (1761–1815), an English chemist. Copper is the predominant metal, but zinc, mercury, silver and/or lead can also be present. Tennantite is the most common and industrially important copper-ore mineral.

OXIDES

Compounds that consist of metallic elements combined with oxygen are placed in the class of minerals called oxides. Some metals, such as silver and gold, will not combine chemically with oxygen, and therefore are not included in this group. Lead oxides are very rare, but tin oxides are abundant natural compounds.

Cuprite

Bisbee, Arizona.

Composition Cu_2O • Copper oxide **Classification** Minerals – Oxides. **Color** Red to very dark red. **Luster** Adamantine to submetallic or earthy. **Transparency** Translucent to transparent. **Streak** Brick red. **Fracture** Conchoidal. **Crystals** Isometric; usually octahedral but many shapes possible; massive, granular. **Formation** Forms in the oxidized parts of copper deposits in association with native copper, malachite, azurite and oxides of iron. **Hardness** 3.5-4. **Cleavage** Poor. **Notes** Cuprite has been observed in giant masses that weigh several tonnes. Cuprite can also form amazing crystal structures in association with pure native copper and a wide array of other secondary copper minerals. This mineral has also been found in fine hair-like masses, called chalcotrichite.

Cuprite has been found at Bisbee, Arizona, and many other locations around the world.

Hematite

Composition Fe_2O_3 • Iron oxide **Classification** Minerals – Silicates. **Color** Crystals are steel gray to black; earthy and massive forms are red to brownish. **Luster** Metallic, or dull. **Transparency** Opaque. **Streak** Deep red. **Fracture** Uneven. **Crystals** Trigonal; normally earthy or sheet-like masses; also reniform (kidney shaped). **Formation** Forms primarily as a hydrothermal and replacement mineral. **Hardness** 5–6. **Cleavage** None. **Notes** Hematite's common name comes from an ancient Greek word meaning "blood," because it gives rocks a dark reddish tinge. This mineral also makes the planet Mars red! Hematite is the most important ore of iron, containing 70% of the metal. Although this mineral is not normally magnetic it may become magnetic when heated.

Yukon.

Magnetite

Merry Widow Mine, British Columbia.

Composition $FeFe_2O_4$ • Iron oxide **Classification** Minerals – Oxides. **Color** Iron-black. **Luster** Metallic to dull. **Transparency** Opaque. **Streak** Black. **Fracture** Conchoidal. **Crystals** Isometric; normally massive or granular; crystals typically octahedral or dodecahedral. **Formation** Forms in a wide variety of environments, including igneous, metamorphic and sedimentary rocks and sands. **Hardness** 5.5-6.5. **Cleavage** None. **Notes** Magnetite is a very common mineral with over 72% iron content, and is the most important iron ore after hematite. Lodestone is a form of magnetite that forms a natural magnet.

Very good clusters of magnetite crystals have been found in Millard Co., Utah. Crystals of magnetite may be as large as several inches across, but most are smaller—approximately 1" (2.5 cm) is the normal size.

CARBONATES

Carbonates are formed when one or more metals or semimetals join with a carbonate atomic structure, $(CO_3)^{-2}$. This class is one of the more abundant and industrially important groups of minerals in the world.

Calcite

Gallatin County, Montana.

Composition $CaCO_3$ • Calcium carbonate **Classification** Minerals – Carbonates. **Color** Normally white or colorless; when impure, however, it can be brown to black, yellow, red, green or blue. **Luster** Vitreous to pearly or dull luster. **Transparency** Translucent to transparent; fluorescent. **Streak** White to grayish. **Fracture** Subconchoidal. **Crystals** Widely variable, rhombohedrons commonest; twinned crystals very common. **Formation** Forms in a wide variety of environments. Excellent crystals often form in hydrothermal veins, alpine fissures and basalt pockets. **Hardness** 3. **Cleavage** Perfect. **Notes** Calcite is one of the most common minerals in the world. Most calcite forms in masses and aggregates, and it is the main component of limestones and marbles. Over 300 types of calcite crystals have been observed. When calcite is dissolved in water, it is often precipitated from drips in limestone caves as stalagmites. Calcite is an important mineral, used in the manufacture of steel, cement and glass.

Azurite

Azurite coating Malachite, Bisbee, Arizona.

Composition $Cu_3(CO_3)_2(OH)_2$ • Copper carbonate hydroxide **Classification** Minerals – Carbonates. **Color** Azure, dark blue. **Luster** Vitreous or dull. **Transparency** Transparent to opaque. **Streak** Blue. **Fracture** Conchoidal. **Crystals** Monoclinic; usually in well-formed equi-dimensional or tabular crystals; also found as crusts, tufts and earthy masses. **Formation** Forms in the oxidized sections of copper deposits. **Hardness** 3.5-4. **Cleavage** Perfect.

Notes This soft mineral is often associated with malachite (see p. 35) and other similar minerals. The bright blue color of azurite helps in easy identification. The presence of water within azurite makes the color of the copper base bright blue rather than the usual green.

Many outstanding specimens of azurite, along with fine specimens of malachite, have been unearthed in the copper deposits in Arizona and Utah.

Cerussite

Bunker Hill Mine, Kellogg, Idaho.

Composition $PbCO_3$ • Lead carbonate **Classification** Minerals – Carbonates. **Color** Varies from colorless to white, gray, yellow or brown. **Luster** Vitreous or resinous. **Transparency** Translucent to transparent. **Streak** White to colorless. **Fracture** Conchoidal. **Crystals** Orthorhombic; normally forms needles, plates and spikes; star-shaped and chevron-shaped twins are sometimes found; also forms crusts on galena (see p. 23). **Formation** Forms in the oxidized sections of mineral veins with lead, copper and zinc. **Hardness** 3-3.5. **Cleavage** Good.

Notes Cerussite is an ore of lead that sometimes fluoresces in ultraviolet light. It produces several crystal shapes, including the spoked star, snowflake and elbow or chevron shapes, which are truly spectacular.

Important sites for this mineral in western North America include Leadville, Colorado; Mammoth-St. Amhony Mine, Tiger, Pinal Co., Arizona; Cerro Gordo, Inyo Co., California; and Moyie and the H. B. Mine at Salmo, British Columbia.

Dolomite (Mineral)

Mt. Brussilof mine, Radium Hot Springs, British Columbia.

Composition $CaMg(CO_3)_2$ • Calcium magnesium carbonate **Classification** Minerals – Carbonates. **Color** Colorless, white, pinkish or light-colored tints. **Luster** Glassy to pearly. **Transparency** Translucent to transparent. **Streak** White. **Fracture** Subconchoidal. **Crystals** Trigonal; normally massive habit; can be prismatic or rhombohedral; also forms saddle-shaped crystals. **Formation** Forms in hydrothermal veins as well as in magnesian limestones. **Hardness** 3.5-4. **Cleavage** Perfect.

Notes Huge beds of the dolomite (mineral) hundreds of feet thick, referred to as dolomitic limestone, have been found around the world. Dolomite looks very similar to calcite, so a specimen should be tested in dilute acid. If it is dolomite, the solution will not fizz, as it does with calcite. Excellent crystals of dolomite have been found in vugs (small to medium-sized cavities in rock) in gold-bearing quartz veins of the Mother Lode district (W. Sierra Nevada) of California.

Magnesite

Mt. Brussiloff mine, Radium Hot Springs, British Columbia.

Composition $MgCO_3$ • Magnesium carbonate **Classification** Minerals – Carbonates. **Color** Varies from colorless to white, gray, yellowish or brown. **Luster** Vitreous to dull. **Transparency** Transparent to translucent. **Streak** White. **Fracture** Conchoidal to uneven. **Crystals** Trigonal & hexagonal; massive habit is common; rhombohedral crystals (rare); also compact and granular habits. **Formation** Forms in hydrothermal veins, metamorphic rocks and sediments. **Hardness** 3.5-4.5. **Cleavage** Perfect.

Notes Magnesite rarely forms crystals. Industrial magnesite comes from very fine-grained massive deposits. Looking something like porcelain, it sticks to the tongue when licked, because the crystals contain tiny holes. Magnesite is also used in making rubber and fertilizers.

Magnesite has been mined in large quantities in Washington and Coast Ranges, California. Massive deposits have also been discovered at Gabbs, Nevada, and on the Yukon River in the Yukon Territory.

Malachite

Bisbee, Arizona.

Composition $Cu_2(CO_3)(OH)_2$ • Copper carbonate hydroxide **Classification** Minerals – Carbonates. **Color** Emerald-green, grass-green, dark green. **Luster** Adamantine, silky, dull. **Transparency** Translucent to opaque. **Streak** Light green. **Fracture** Subconchoidal to uneven. **Crystals** Monoclinic; massive forms are botryoidal (shaped like a bunch of grapes), stalactitic or globular. Crystals are shaped like tufts of needles. **Formation** Forms in the oxidized portions of copper deposits. **Hardness** 3.5-4. **Cleavage** Perfect. **Notes** Because malachite is beautiful and somewhat soft, it has been carved into ornaments and worn as jewelry for thousands of years. The bright green color is so distinctive, it is easy to identify, especially if the specimen shows concentric internal banding.

Malachite is a secondary copper mineral. Fine malachite specimens in western North America have come from copper mines in California, Nevada and Utah.

Rhodochrosite

Emma Mine, Montana.

Composition $MnCO_3$ • Manganese carbonate **Classification** Minerals – Carbonates. **Color** Normally pink to red, but may also be brown, orange or yellowish; darkens on exposure to the atmosphere. **Luster** Vitreous to pearly. **Transparency** Translucent to transparent. **Streak** White. **Fracture** Uneven. **Crystals** Trigonal; normally massive, scalenohedrons, globules, fills veins and nodules or forms stalactites; occasionally forms rhombohedrons or prismatic crystals. **Formation** Forms in hydrothermal veins and in altered manganese deposits. **Hardness** 3.5-4. **Cleavage** Perfect. **Notes** Rhodochrosite is easily recognizable by its beautiful rose-pink color, but good specimens are very uncommon. Excellent crystals have been discovered in several mines in Colorado, where the mineral is a deep pink. Its name originates from the Greek rhodon, meaning "rose." and chros, meaning "color."

BORATES

Borates are formed when metallic elements combine with the borate $(BO_3)^{-3}$ radical. Some of the artificial boron compounds are among the hardest substances known to man.

Borax

Composition $Na_2B_4O5(OH)_4.8H_2O$ • Hydrated sodium borate
Classification Minerals – Borates. **Color** Deep blue to sky-blue, greenish blue.
Luster Vitreous, dull. **Transparency** Transparent to translucent, but crystals lose water and turn white. **Streak** White. **Fracture** Conchoidal. **Crystals** Triclinic; normally short prismatic, also thick tabular, incrusting, stalactitic. **Formation** Forms around hot springs and in evaporite deposits. **Hardness** 2.5. **Cleavage** Perfect.
Notes Some of the world's largest borax deposits are mined in California and Nevada. The crystals can be quite large—single crystals reaching 6 or 7" (15-20 cm) long and 1 pound (450 g) in weight have been recorded.

Caution Borax is a poisonous mineral. Be sure to wash your hands after contact.

California.

SULFATES, TUNGSTATES & MOLYBDATES

Sulfates are a group of minerals comprising one or more metals along with a sulfate compound: $(SO_4)^{-2}$. They are produced when sulfides are exposed to air. All members of this group are soft, light-colored and often translucent or transparent.

Molybdates and tungstates are heavy, soft and brittle minerals that have the same chemical structure as sulfates, except that tungsten and molybdenum replace the sulfur in the basic compound.

Barite

Barite on Fluorite, Rock Candy Mine, British Columbia.

Composition $BaSO_4$ • Barium sulfate **Classification** Minerals – Sulfates. **Color** White, gray or colorless, also various shades of yellow, brown, red or blue. **Luster** Vitreous, pearly. **Transparency** Transparent to translucent. **Streak** White. **Fracture** Uneven. **Crystals** Orthorhombic; primarily tablet and prism shapes; also gains, plates, cockscombs and rosettes. **Formation** Forms in hydrothermal deposits. **Hardness** 3–3.5. **Cleavage** Perfect.

Notes Barite is abnormally heavy for a non-metallic mineral. Its high density makes it very useful in industry. Barite is used in drilling oil wells. Powdered barite is used to increase the bulk in white paper and give it a fine, smooth finish. One of the crystals barite forms sometimes grows as thin, blade-like crystals that may cluster into a cockscomb-like formation. Some varieties of barite are fluorescent.

Chalcanthite

Butte, Montana.

Composition $CuSO_4$-$5H_2O$ • Hydrated copper sulfate **Classification** Minerals – Sulfates. **Color** Sky-blue to dark blue, greenish blue, greenish. **Luster** Vitreous to silky. **Transparency** Translucent to transparent. **Streak** Pale blue to colorless. **Fracture** Conchoidal. **Crystals** Triclinic; typically stubby prisms (rare) and thick tablets; primarily as grape-like and stalactite-like masses and crusts. **Formation** Forms in oxidized pans of copper sulfide veins. **Hardness** 2.5 **Cleavage** Poor.

Notes Chalcanthite is found most frequently in areas with an arid climate. In western North America, noteworthy sites include Imlay, Pershing Co., Nevada; and Bisbee, Cochise Co., Ajo, Arizona.

Caution Chalcanthite is a poisonous mineral. Be sure to wash your hands after contact.

Cyanotrichite

Lemhi County, Idaho.

Composition $Cu_4Al_2(SO_4)(OH)_{12}.2H_2O$ • Hydrated copper aluminum sulfate hydroxide. **Classification** Minerals – Sulfates **Color** Sky-blue to azure-blue. **Luster** Vitreous to silky. **Transparency** Transparent to translucent. **Streak** Pale blue. **Fracture** Uneven. **Crystals** Orthorhombic; crusts, radial sprays and balls of minute needle-like crystals. **Formation** Forms in the oxidized zone of ore veins, especially copper. **Hardness** 3. **Cleavage** None.

Notes Cyanotrichite is truly one of the most strikingly colored of all minerals. Its name originates from its deep blue color, caused by the presence of water. The best field marks for cyanotrichite are the radiating sprays or balls of fine crystals and the distinctive color.

Linarite

Darwin Mine, California.

Composition $PbCu(SO_4)(OH)_2$ • Lead copper sulfate hydroxide **Classification** Minerals – Sulfates. **Color** Deep azure-blue. **Luster** Vitreous to subadamantine. **Transparency** Transparent to translucent. **Streak** Pale blue. **Fracture** Conchoidal. **Crystals** Monoclinic; elongated prismatic crystals that are often tabular. **Formation** Forms in oxidized lead and copper veins. **Hardness** 2.5. **Cleavage** Perfect.

Notes Under normal circumstances, linarite forms as a coating of minute, bright blue crystals, but it can also form large masses in association with other copper minerals.

Superb crystals reaching 4" (10 cm) in length have been unearthed in the Mammoth Mine at Tiger, Pinal Co., Arizona.

Scheelite

Anaconda mine, British Columbia.

Composition $CaWO_4$ • Calcium tungstate **Classification** Minerals – Tungstates **Color** Variable from white to colorless, gray, yellowish, orange, brownish and greenish. **Luster** Vitreous to adamantine. **Transparency** Transparent to translucent. **Streak** White. **Fracture** Subconchoidal to uneven. **Crystals** Tetragonal; usually dipyramidal, also incrusting, granular or compact. **Formation** Forms in hydrothermal veins, in contact metamorphic rocks and in pegmatites. **Hardness** 4.5–5. **Cleavage** Distinct.

Notes Scheelite, an important ore of tungsten, is a beautiful mineral with a bright blue-white fluorescence under shortwave ultraviolet light. This property has helped miners discover crystals in the dark using ultraviolet lamps.

Wulfenite

Composition $PbMoO_4$ • Lead molybdate **Classification** Minerals – Molybdates. **Color** Brilliant orange to yellow, as well as brown, gray or greenish brown. **Luster** Resinous to adamantine. **Transparency** Transparent to translucent. **Streak** White. **Fracture** Subconchoidal to uneven. **Crystals** Triclinic; variable, especially square tabular; also extremely thin; cube-like, massive or granular. **Formation** Forms in ore veins that have been altered by circulating fluids, especially water. **Hardness** 3. **Cleavage** Distinct.

Mammoth Mine, Tiger, Arizona.

Notes Wulfenite, a minor ore of lead, is known for its brilliant, colorful and varied crystals. Some of the brightest orange wulfenite crystals originate from Red Cloud Mine, Yuma, Arizona.

PHOSPHATES, ARSENATES AND VANADATES

This large class of minerals includes compounds of metallic elements with phosphate $(PO4)^{-8}$; arsenate $(ASO4)^{-8}$, $(ASO_3)^{-1}$; or vanadate $(VO_4)^{-3}$ $(VO_3)^{-1}$ radicals. Although their properties are variable, they tend to be very colorful, soft, brittle and well crystallized.

Pyromorphite

Bunker Hill Mine, Kellogg, Idaho.

Composition $Pb_5(PO_4)_3Cl$ • Lead chlorophosphate **Classification** Minerals – Phosphates. **Color** Yellow-green. **Luster** Resinous to greasy. **Transparency** Transparent to translucent. **Streak** White to pale yellow or greenish yellow. **Fracture** Uneven to subconchoidal. **Crystals** Trigonal, hexagonal; slender prism-shaped, often hollow-ended. **Formation** Forms as a secondary mineral in the oxidation zone of lead veins. **Hardness** 3.5-4. **Cleavage** Imperfect. **Notes** Pyromorphite, a minor ore of lead, typically forms crusts and masses, but it is best known for its striking green crystals. Excellent examples of pyromorphite occur on the Society Girl claim, steel mining district in British Columbia. Striking drusy green crystals occur at several locations in the Coeur d'Alene district, Shoshone Co., Idaho.

Variscite

Turquoise Mountain Mine, Idaho.

Composition $Al(PO_4).2H_2O$ • Hydrated phosphate of aluminum and iron **Classification** Minerals – Phosphates. **Color** Pale green to emerald-green, also bluish green and colorless. **Luster** Vitreous to waxy. **Transparency** Translucent to opaque. **Streak** White. **Fracture** Uneven, splintery and conchoidal. **Crystals** Orthorhombic; crystals (very rare); normally as fine-grained masses, nodules and crusts. **Formation** Forms at sites where phosphate-rich water has altered aluminum-rich rocks. **Hardness** 3.5–4.5. **Cleavage** Perfect. **Notes** In North America, variscite is found only in the western United States. Some of the finest specimens originate from Clay Canyon, Utah Co., Utah, where masses over 3.3' (1 m) in diameter have been unearthed. Other known locations include Esmeralda and Mineral Cos., Nevada.

Vivianite

Blackbird Mine, Lemhi County, Idaho.

Composition $Fe_3(PO_4)_2.8H_2O$ • Hydrated iron phosphate **Classification** Minerals – Phosphates. **Color** Colorless to green or blue to indigo, darkening on exposure to light. **Luster** Vitreous. **Transparency** Transparent to translucent. **Streak** White or bluish green. **Fracture** Uneven. **Crystals** Monoclinic; normally in short to long prismatic crystals; also radial fibrous. **Formation** Forms within the oxidation zone of iron- and manganese-rich deposits. **Hardness** 1.5–2. **Cleavage** Perfect.

Notes When vivianite first forms, it is almost colorless, but when it is exposed to light it turns blue, and then gradually completely black. It also becomes brittle. It is therefore recommended that collectors keep specimens in a dark place. Beautiful blue crystals of vivianite have been found in Idaho, Utah and Colorado.

Vanadinite

Apache Mine, Globe, Arizona.

Composition $Pb_5(VO_4)_3Cl$ • Lead chloro-vanadate **Classification** Minerals – Vanadates. **Color** Red to brown, brownish yellow or orange. **Luster** Resinous to subadamantine. **Transparency** Transparent to translucent. **Streak** White or yellowish. **Fracture** Conchoidal. **Crystals** Hexagonal; normally small prismatic crystals, sometimes hollow; also globular. **Formation** Forms in the oxidation zone of lead deposits. **Hardness** 3. **Cleavage** None. **Notes** Vanadinite is a minor ore of lead. Its name is derived from the mineral's composition. It normally forms in dry and desert regions, such as the southwest USA.

Amazing red crystals have been found at the Apache Mine, Globe district, Gila Co., Arizona. Specimens have also been found in the El Dorado Mine, near Indio, Riverside Co., California.

SILICATES

The silicates make up the largest and most common class of minerals. All silicates contain the silicate atomic structure $(SiO_4)^{-4}$. This is the fundamental building block, the tetrahedron. In this basic compound, one silicon atom is surrounded by four equally spaced oxygen atoms.

Silicates are divided into six classes based upon the number and manner of linking silicon-oxygen tetrahedrons. Metallic elements combine with either single or linked Si-O structures to make up the various silicate minerals.

Actinolite Nephrite (Jade)

Composition $Ca_2(Mg,Fe)_5Si_8O_{22}(OH)_2$ Calcium magnesium iron silicate hydroxide **Classification** Minerals – Silicates. **Color** Varies greatly from white to green to violet. **Luster** Vitreous, pearly or silky. **Transparency** Transparent to translucent. **Streak** Colorless. **Fracture** Splintery and uneven. **Crystals** Monoclinic; normally long prismatic with a diamond-shaped cross-section. Often found in massive aggregates. **Formation** Often forms in amphibolites from metamorphic rocks and massive hydrothermal replacement deposits. Amphibolites are minerals that normally form long, wedge-shaped crystals. **Hardness** 5-6. **Cleavage** Perfect. **Notes** Actinolite is a very tough mineral made up of interlocking

Polar Jade, Jade West Mines, Lillooet Jade Walk, British Columbia.

fibrous minerals that are visible only under the microscope.

One type of actinolite is the smooth green nephrite, one of two "jades." Jadeite is the other (see below.) Nephrite is usually a chemical mixture of actinolite and tremolite (see p. 62). Significant quantities have been found in British Columbia, Alaska, California and Wyoming.

Twin Rivers Quarry, Lake Wenatchee, Chelan County, Washington.

Wenatchee Ridge, Washington.

Jadeite

Jadeite is the second mineral that is included in the term "jade." This is a harder mineral that commands a much higher price. It is rare and is of gem quality. It has been found in San Benito County, California, and other locations around the globe.

Chalcedony

Polished Yellow Agates, Oregon.

Composition SiO_2 • Silicon dioxide
Classification Minerals – Silicates. **Color** Varies from white to black and all the colors of the rainbow. **Luster** Vitreous or resinous. **Transparency** Transparent to translucent. **Streak** White. **Fracture** Brittle and conchoidal. **Crystals** None; chalcedony is a microscopic variety of quartz (cryptocrystalline). **Formation** Formed in several situations, normally near the surface of the Earth with relatively low temperatures and pressures. **Hardness** 7. **Cleavage** None.
Notes Chalcedony (pronounced kal-sed′-o-ny) is a form of quartz that is banded with a wide variety of lines or patterns. Chalcedonies are believed to have been the earliest hard materials used by mankind. Arrowheads, knives, tools, cups and bowls were all shaped from various types of chalcedony.

Polished Black Agates, Oregon.

Agate is a translucent type of chalcedony that often has various bands or other patterns. The agate is the birthstone for May. Among the many varieties of agate are:
Blue lace agate: wavy bands of mauve-blue and white.
Fire agate: limonite inclusions.
Thundereggs: a coarse outer covering encasing an agate-filled center. Eastern Oregon is famous for its thundereggs.
Scenic agate: resembling a woodland scene.
Montana agate: clear matrix with orange-red and black inclusions.
Bloodstone: distinctly banded; adjacent bands differ in color and degree of

Montana Agate in the rough, Montana.

Polished Montana Agate, Montana.

translucency.
Jasper: opaque, with impurities that produce a rainbow of colors.
Sugar agate: white, opaque and not very interesting.
White agate: milky or semi-transparent, lacking impurities that would generate other colors.
Cloud agate: semi-translucent and cloudy-looking.
Banded agate: having bands of alternating colors.
Black agate: black, gray or dark blue and normally opaque.
Jasper: opaque, red or green.

Sugar Agate, British Columbia.

Polished Jasper, British Columbia.

Traces of iron oxide give jasper its red hue. The green color comes from microscopic fibers of actinolite or chlorite in the rock. Jasper pebbles are often found on beaches in areas of igneous rock. They glisten when wet and they take polishing well. In western North America, jasper is found more often in Oregon, Arizona and California than in other areas.

Most of these materials are favored by makers of cabochons and cameos.

Banded Agate, Tulameen, British Columbia.

Fortification Agates, Oregon.

Almandine (Garnet)

Composition $Fe_3^{+2}Al_2(SiO_4)_3$ • Iron aluminum silicate
Classification Minerals – Silicates. **Color** Deep red to brown or brownish black. **Luster** Vitreous. **Transparency** Transparent to translucent. **Streak** White. **Fracture** Subconchoidal. **Crystals** Isometric; dodecahedrons and trapezohedrons are very common; also granular, lamellar and compact. **Formation** Forms in regionally metamorphosed rocks, including schist. **Hardness** 6.5-7.5. **Cleavage** None.
Notes Almandine is sometimes referred to as the iron-garnet. Garnets are a group of very closely related aluminum silicates and calcium silicates. These dense minerals are formed deep within the Earth, only under conditions of intense heat and pressure. Almandine is considered a gemstone. Some garnets are of igneous origin, but most are products of metamorphism.

Well-formed crystals of almandine have been recovered from Wrangell Island, Alaska, and Chaffee County, Colorado. Gem-quality almandine is found in Alaska and Idaho.

Almandine in Mica Schist, Bella Vista Mine, Mitkof Island, Alaska.

Anorthite

Spectrum Mine, Plush, Lake County, Oregon.

Composition $CaAl_2Si_2O_8$ • Calcium aluminum silicate
Classification Minerals – Silicates. **Color** Colorless, off-white, gray, pinkish or greenish. **Luster** Vitreous. **Transparency** Transparent to translucent. **Streak** White. **Fracture** Uneven. **Crystals** Triclinic; always twinned but crystals are very rare; additional habits are lamellar or massive. **Formation** Forms in many igneous rocks, especially those with basic composition. **Hardness** 6-6.5. **Cleavage** Perfect.
Notes Anorthite is a common mineral in basic rocks that form at high temperatures, such as gabbros. It is also present in meteorites and moon rocks! Noteworthy deposits of anthorite have also been found in the Aleutian Islands, Alaska, as well as Lake Co. and Nevada Co., California. Masses of anorthite have been collected from a body of serpentinite (see. p. 87), near Middletown, Lake Co., California.

Anthophyllite

Scotch Creek, British Columbia.

Composition $(Mg,Fe)_7Si_8)O_{22}(OH)_2$ • Magnesium iron silicate hydroxide **Classification** Minerals – Silicates. **Color** Brown to gray. **Luster** Vitreous or silky. **Transparency** Transparent to nearly opaque. **Streak** Gray. **Fracture** Uneven. **Crystals** Orthorhombic; prismatic but rare. **Formation** Forms in crystalline schists and gneisses, and from metamorphic rocks. **Hardness** 5.5-6. **Cleavage** Perfect. **Notes** Anthophyllite is easily identified by its color and its fibrous masses. Although well-formed crystals of anthophyllite are quite rare, some crystalline aggregates can be impressive.

Amazing blue schiller reflections characterize an unusual anthophyllite amphibolite located near Butte, Montana. **Caution** Anthophyllite often contains asbestos fibres, which are a health hazard. Keep the mineral away from your nose and mouth, and wash your hands carefully.

Beryl

Near Superior, Montana.

Composition $Be_3Al_2Si_6O_{18}$ • Beryllium aluminum silicate **Classification** Minerals – Silicates. **Color** Green to yellow, gold. red, or pink, or may be colorless. **Luster** Vitreous. **Transparency** Transparent to translucent. **Streak** White to colorless. **Fracture** Uneven to conchoidal. **Crystals** Trigonal; hexagonal; also in massive, compact and columnar habits. **Formation** Forms in pegmatites (see p. 72) and granites, and in some metamorphosed rocks. **Hardness** 7.5-8. **Cleavage** Poor. **Notes** Beryl is a very hard mineral. Its prismatic crystals are sometimes terminated with small pyramids. Often the crystals are striated parallel to their length, and can reach huge sizes. Specimens that extend 18 feet (5.5 m) long have been discovered.

Some excellent aquamarine crystals have been found on pegmatites (see p. 72) on Mt. Antero in Chaffee Co., Colorado. Other fine sites for crystals include Mesa Grande and Pala, San Diego Co., California. Pale-green betyl occurs in a pegmatite near Oreana, Pershing Co., Nevada.

Chabazite

Composition $Ca_2(Al_4Si_8O_{24}).13H_2O$ • Hydrated calcium potassium sodium aluminum silicate

Classification Minerals – Silicates. **Color** Normally white, but can be yellow, pink or red. **Luster** Vitreous. **Transparency** Transparent to translucent. **Streak** White. **Fracture** Uneven. **Crystals** Rhombohedral, but resemble cubes. **Formation** Forms in the cavities of basaltic lavas and in some limestones. Also forms small crystals around hot springs. **Hardness** 4-5. **Cleavage** Poor.

Notes Chabazite is a member of the zeolite group of minerals, 50 or so related silicate minerals that commonly form in cavities in basic igneous rocks, especially vesicular basalt. They are popular with collectors because of their beauty. The name zeolite is a Greek reference to the ease in which these minerals boil in fusing under a flame.

Excellent translucent crystals have been found with other zeolites near Goble, Columbia Co., Oregon.

Chabazite with green clay inclusions, Oxbow Dam, Oregon

Chrysotile

Composition $Mg_3Si_2O_5(OH)_4$ • Hydrous magnesium silicate

Classification Minerals – Silicates. **Color** Grayish to white, or green, yellow or brown. **Luster** Silky to greasy. **Transparency** Translucent. **Streak** White. **Fracture** Splintery. **Crystals** Monoclinic; normally massive. **Formation** The serpentine minerals are formed by hydrothermal metamorphic processes. **Hardness** 2.5. **Cleavage** None.

Cassiar, British Columbia.

Notes Chrysotile is a member of the serpentine group of silicate minerals (see p. 58). The name "serpentine" refers to the serpent-like surface pattern of serpentinite rocks. Antigorite is another similar mineral in the serpentine group. In western North America, chrysotile asbestos was once mined near Copperopolis, Calaveras Co., California. Chrysotile asbestos is used for many thermal or electrical insulation applications, including fireproof fabrics and brake linings.

Caution Chrysotile often contains asbestos fibers, which are a health hazard. Keep this mineral away from your nose and mouth, and wash your hands carefully.

Clinochlore

Composition $(Mg,Fe^{+2})_5Al(Si,Al)_4.O_{10}(OH)_8$ • Hydrous magnesium iron aluminum silicate **Classification** Minerals – Silicates. **Color** White to yellowish, green or colorless. **Luster** Pearly. **Transparency** Transparent to opaque. **Streak** Colorless to greenish white. **Fracture** Uneven. **Crystals** Monoclinic; tabular with hexagonal cross-section, as well as in massive, foliated, scaly and granular habits. **Formation** Found in many chlorite schists, talc schists and various metamorphic rocks. **Hardness** 2-2.5, **Cleavage** Perfect.

Calaveras County, California.

Notes Clinochlore is a soft silicate that flakes when it is heated, rather than fusing. It is one of the rarer minerals found in western North America.

Elbaite Toumaline

Composition $Na(Li,Al)_3Al_6Si_6O_{18}(BO_3)_3,(BO_3)_3.(OH)_4$ • Sodium lithium aluminum bora-silicate hydroxide **Classification** Minerals – Silicates. **Color** Normally black to bluish black, but may be blue, pink or green. **Luster** Vitreous. **Transparency** Transparent to opaque. **Streak** Colorless. **Fracture** Uneven. **Crystals** Trigonal; normally elongated triangular or hexagonal (six-sided) prisms, with surface striations. **Formation** Forms in granites and pegmatites, as well as in metamorphic rocks. **Hardness** 7-7.5. **Cleavage** None.

Himalaya Mine, California.

Notes Tourmaline can be any color of the rainbow—in fact, it is one of the most variably colored gems in the world. The ancient Egyptians believed it gathered all the colors of the rainbow as it made its way up through the Earth. They referred to it as the rainbow rock.

Truly fine red, blue and green tourmaline crystals have been found in pegmatites at Riverside and San Diego Cos. in California.

Epidote

Calvert Creek Mine, Montana.

Composition $Ca_2(Al,Fe)_3(SiO_4)_3(OH)$ • Calcium aluminum iron silicate hydroxide **Classification** Minerals – Silicates. **Color** Primarily pistachio or dark greenish black; other colors include yellowish green to brownish black. **Luster** Vitreous. **Transparency** Transparent to nearly opaque. **Streak** White to grey. **Fracture** Uneven. **Crystals** Monoclinic; usually in long, slender, grooved prismatic crystals; also forms crusts, masses and grains. **Formation** Forms in both metamorphic and igneous rocks. **Hardness** 6-7. **Cleavage** Perfect.

Notes Beautiful epicote crystals have been unearthed from several North American sites. Both groups and single crystals have come from locations near Sulzer, Prince of Wales Island, Alaska. Amazing crystals up to 1.5" (3.75 cm) across were encountered in the Calumet iron mine, Chaffee Co., Colorado, and huge crystals, reaching 12" (30 cm) in length, were found in the Seven Devils district, Adams Co., Idaho. Striking pale green crystals up to 6" (15 cm) in length have been found at several small tungsten mines in the Greenhorn Mts., Kern Co., California.

Hemimorphite

Composition $Zn_4Si_2O_7(OH)_2.H_2O$ • Hydrous zinc silicate **Classification** Minerals – Silicates. **Color** White, colorless, yellowish, bluish or brownish. **Luster** Vitreous or silky. **Transparency** Transparent to translucent. **Streak** Colorless. **Fracture** Uneven to conchoidal. **Crystals** Orthorhombic; normally thin tabular and in fan-shaped aggregates; also in coxcomb groups, mammillary and stalactitic. **Formation** Forms where zinc veins have been oxidized. **Hardness** 4.5-5. **Cleavage** Perfect.

Elkhorn, Montana.

Notes The blade-shaped crystals and radiating aggregates are the best field marks for this mineral. However, hemimorphite is a mineral of many forms. It also occurs as massive, compact, granular, botryoidal, stalactitic, fibrous and encrusting habits. Hemimorphite is a minor ore of zinc.

Heulandite

Composition $(Ca,Na)_{2-3}Al_3(Al,Si)_2Si_{13}O_{36}.12H_2O$ • Hydrated calcium sodium aluminum silicate
Classification Minerals – Silicates. **Color** White, pink, red or brown. **Luster** Vitreous to pearly. **Transparency** Transparent to translucent. **Streak** White. **Cleavage** Poor. **Fracture** Uneven. **Crystals** Monoclinic; usually in coffin-shaped or tabular single crystals. **Formation** Forms in cavities of basalts, usually associated with calcite and other zeolites. **Hardness** 3.5-4. **Cleavage** Perfect.
Notes Heulandite is a low-pressure mineral in the zeolite group of silicates. The zeolites are a large family of minerals that are related in composition, in occurrence and many in appearance. Zeolite comes from the Greek, referring to the ease with which these minerals boil in fusing under a flame.

Heulandite is often found in association with other zeolites. In western North America it has been found in Idaho, Oregon and Washington. Pink and white crystals have been found in basalt cavities near Edwards, Tillamook Co., Oregon.

Heulandite on Mordenite, Rat's Nest Mine, Idaho.

Ilvaite

Composition $CaFe_2^{+2}Fe^{+3}(SiO_4)_2(OH)$ • Hydrous calcium iron silicate
Classification Minerals – Silicates. **Color** Noticeably dark; black to grayish brown or brownish black. **Luster** Resinous to vitreous. **Transparency** Opaque. **Streak** Black, often with greenish or brownish tints. **Cleavage** Distinct. **Fracture** Uneven. **Crystals** Orthorhombic and monoclinic; elongated prismatic with striations on the prism faces. **Formation** Forms in rocks that have been intruded by magma or in contact with lava (contact metamorphism). **Hardness** 5.5-6. **Cleavage** Distinct.
Notes Ilvaite occurs in several ways, including crystal faces that are striated vertically, massive, columnar and compact habits. Ilvaite fuses very easily in a flame, producing a black magnetic globule.

Lacey Mine, South Mountain, Idaho.

Kyanite

Composition Al_2SiO_5 • Aluminum silicate
Classification Minerals – Silicates. **Color** White to almost black, but more often blue, gray, green, yellow or pink. **Luster** Vitreous. **Transparency** Transparent to translucent. **Streak** Colorless. **Cleavage** Perfect. **Fracture** Splintery. **Crystals** Triclinic; normally in bladed crystal aggregates, also in terminated crystals. **Formation** Forms in conditions of high temperatures and pressures. **Hardness** Varies: lengthwise 4-5; crosswise 6-7. **Cleavage** Perfect.
Notes The distinctive bluish blades of kyanite are its most common color. In California, large minable masses of kyanite were discovered in the Cargo Muchacho Mts., Imperial Co. Massive forms like these are very useful to humans, because they are so heat-resistant. Kyanite is turned into a fiber and used to make a range of products, from insulators for car spark plugs to containers for molten steel.

Ennis, Montana.

Laumontite

Pine Creek Mine, California.

Composition $CaAl_2Si_4O_{12}.4H_2O$ • Hydrous calcium aluminum silicate **Classification** Minerals – Silicates. **Color** Colorless, or white, yellowish or brownish. **Luster** Vitreous to pearly. **Transparency** Transparent to opaque. **Streak** Colorless. **Cleavage** Perfect. **Fracture** Uneven. **Crystals** Monoclinic; normally short prismatic crystals with oblique terminations; massive habit as well. **Formation** Forms in cavities in basalt of volcanic rocks, along with calcite and zeolites, at low temperatures. **Hardness** 3-4. **Cleavage** Perfect.
Notes Laumontite has a white powdery surface because of the loss of water within. Long, prismatic crystals reaching 3" (7.5 cm) in length have been discovered.

This mineral was named in honor of Gillet de Laumont (1747-1834), a French mineralogist who discovered the mineral in the lead mines at Brittany in 1785.

Lepidolite

Composition $K(Li,Al)_3(Si,Al)_4O_{10}(F,OH)_2$ • Potassium lithium aluminum silicate hydroxide fluoride
Classification Minerals – Silicates. **Color** Pink to purple, grayish to white, or may be colorless. **Luster** Pearly. **Transparency** Transparent to translucent. **Streak** Colorless to white. **Cleavage** Perfect basal. **Fracture** Uneven. **Crystals** Monoclinic; usually in scaly aggregates: well-developed crystals (rare); also foliated and compact. **Formation** Found at granite pegmatites, as fine-granular masses or as stubby or tabular crystals in cavities. **Hardness** 2.5-3. **Cleavage** Perfect.

Notes Western North America hosts lepidolite at several locations, where mineral veins are rich in tin. Mines with fine masses have been established at the Stewart Pegmatite Mine, San Diego Co., California, and amazing crystals have been obtained from the Little Three Pegmatite in San Diego Co., California. Substantial amounts have also been mined in the Black Hills of South Dakota and in southern California.

San Diego County, California.

Muscovite

Composition $KAl_2(Si_3Al)O_{10}(OH,F)_2$ • Potassium aluminum silicate hydroxide fluoride
Classification Minerals – Silicates. **Color** White or colorless to yellowish, greenish, pink, brownish or multicolored. **Luster** Vitreous to pearly. **Transparency** Transparent to translucent. **Streak** Colorless. **Cleavage** Perfect basal. **Fracture** Uneven. **Crystals** Monoclinic; tabular, often hexagonal in outline; also foliated, flaky, scaly to fibrous. **Formation** Forms in igneous rocks, especially those of acid composition, such as granite, and in metamorphic rocks. **Hardness** 2-2.5. **Cleavage** Perfect.

South of Hamilton, Montana.

Notes The thin flakes of muscovite clearly identify it as a mica. The best muscovite specimens are only found in granitic pegmatites, which exist throughout North America.

Muscovite is very heat-resistant and is therefore used for windows in stoves. It is also an amazing electrical insulator, and it makes wonderful artificial snow for Christmas trees. Thin sheets of transparent muscovite (sometimes called isinglass) are still used as windows in iron stoves. Muscovite is also used as an insulator in electrical equipment.

Natrolite

Natrolite with Apophyllite, Weyerhaeuser Lincoln Cr. Quarry, Washington.

Composition $Na_2(Al_2Si_3)O_{10}.2H_2O$ • Hydrous sodium aluminum silicate **Classification** Minerals – Silicates. **Color** White, colorless, grayish, yellowish or reddish. **Luster** Vitreous to pearly. **Transparency** Transparent to translucent. **Streak** Colorless. **Cleavage** Perfect. **Fracture** Uneven. **Crystals** Orthorhombic; slender to acicular (needle-shaped); also radial, granular, compact; commonly lines cavities in volcanic rocks. **Formation** Lines the cavities in basalts and other lavas. **Hardness** 5-5.5. **Cleavage** Perfect.

Notes Natrolite loses water at around 300°C (572°F) and then reabsorbs it quickly from atmospheric humidity. Crystals have been discovered up to 3" (7.5 cm) in length near Livingston, Montana. Natrolite has been found in Colorado, and in Oregon traprock areas where large crystals reaching 8" (12 cm) in length have been found.

Natrolite is one of three very similar silicate minerals. Scolecite is a calcium equivalent $[Ca(Al_2Si_3)O_{10}.3H_2O]$ that forms large sprays. Mesolite has an intermediate composition, with both Ca and Na in its chemistry $[Na_2Ca_2(Al_6Si_9)O_{30}.8H_2O]$. The most common form of mesolite is clusters of fine, silky-white needle-like crystals.

Neptunite

Composition $KNa_2Li(Fe^{+2},Mn^{+2})_2Ti_2Si_8O_{24}$ • Sodium potassium iron titanium silicate **Classification** Minerals – Silicates. **Color** Black, with red-brown internal reflections. **Luster** Vitreous. **Transparency** Translucent to opaque. **Streak** Black, with some internal reddish brown reflections. **Cleavage** Perfect. **Fracture** Conchoidal. **Crystals** Monoclinic; normally well-formed prismatic crystals with square cross-section. **Formation** Forms as an accessory mineral in some igneous rocks. **Hardness** 5-6. **Cleavage** Perfect.

Neptunite (black) on Natrolite, California.

Notes This mineral was discovered in California in the 1900s. Here it occurs with benitoite and natrolite near the headwaters of the San Benito River, San Benito Co. Neptunite was named after Neptunus, the Roman god of the sea.

Neptunite is often found in association with natrolite and benitoite.

Orthoclase Feldspar

Colorado.

Composition $KAlSi_3O_8$ • Potassium aluminum silicate
Classification Minerals – Silicates. **Color** Off-white; occasionally yellow or reddish. **Luster** Vitreous. **Transparency** Transparent to translucent. **Streak** White. **Cleavage** Perfect. **Fracture** Uneven to conchoidal. **Crystals** Monoclinic; normally as single crystals; twin crystals common; also cleavage masses, granular. **Formation** Forms in many igneous and metamorphic rocks. **Hardness** 6-6.5. **Cleavage** Perfect.
Notes To identify orthoclase, look for blocky cleavage and striations on the surface. This common raw material is used to make porcelain products.

Although orthoclase occurs in a great many places around the world, some noteworthy sites are found in western North America. White adularia (with a bluish sheen) crystals up to 1" (2.5 cm) across have been discovered in gold-bearing quartz veins at Bodie, Mono Co., California, and in silver mines in the Silver City district, Owhyee Co., Idaho.

Phlogopite

Composition $KMg_3AlSi_3O_{10}(OH)_2$ • Potassium magnesium aluminum silicate hydroxide **Classification** Minerals – Silicates. **Color** Brown to reddish brown or coppery. **Luster** Vitreous to pearly. **Transparency** Transparent to translucent. **Streak** White. **Cleavage** Perfect in one direction. **Fracture** Uneven. **Crystals** Monoclinic; often prismatic crystals with hexagonal outline; also plates, scales. **Formation** Found primarily in metamorphosed limestones and ultrabasic rocks such as pyroxenites and peridotites. **Hardness** 2.5-3. **Cleavage** Perfect.
Notes Phlogopite is soluble in concentrated sulfuric acid, which is one test for correct identification of this and other minerals. The similar muscovite does not decompose in sulfuric acid. Phlogopite crystals as large as 1" (2.5 cm) across have been found in East Fresno Co., California.

Burgess, British Columbia.

Quartz

Quartz, variety Amethyst, Lake Creek, Oregon.

Composition SiO_2 • Silicon dioxide
Classification Minerals – Silicates. **Color** Quite variable but translucent is very common; may be cloudy or milky, purple or smoky brown. **Luster** Vitreous or resinous. **Transparency** Normally translucent to opaque; transparent when colorless. **Streak** Colorless to white. **Cleavage** Poor. **Fracture** Conchoidal. **Crystals** Trigonal; variable, commonly hexagonal prisms ending in six-sided pyramids. **Formation** Quartz is the most common mineral in the Earth's crust. It occurs in igneous, metamorphic and sedimentary rocks, and is frequently found in mineral veins with metal ores. **Hardness** 7. **Cleavage** None. **Notes** Quartz is a tough mineral that doesn't break down easily. Crystallized quartz occurs in several different colors. Violet crystals are referred to as amethyst; gray to black crystals are called smoky quartz. These and other hues are caused by impurities.

Quartz has many uses. The sand is used in manufacturing glass and in making fused silica. The clear rock crystal is in great demand for electronic equipment, such as oscillators for controlling radio frequencies. Colored varieties are also very desirable in the gem trade.

Drusy Quartz, Grand Forks, British Columbia.

Quartz was formerly considered to be an oxide, but because of its internal crystal structure and its physical properties, it is now categorized with the silicates.

Drusy Quartz is a term to describe a cluster of many tiny or partially formed quartz crystals. Chalcedony (see p. 44) is a type of quartz.

Quartz, hollow Amethyst egg within an egg, Willow Springs, Utah.

Quartz inside Dryhead Agate, Carbon County, Montana.

Rhodonite

Composition $(Mn^{+2}Fe^{+2}Mg,Ca)SiO_3$ • Manganese silicate
Classification Minerals – Silicates. **Color** Pink to rose-red or brownish red. **Luster** Vitreous. **Transparency** Transparent to translucent. **Streak** Colorless. **Fracture** Conchoidal to uneven. **Crystals** Triclinic; crystals usually large and flat. **Formation** Forms in metamorphic rocks as well as in hydrothermal replacement deposits. **Hardness** 5.5-6. **Cleavage** Perfect.

Rhodonite & Quartz, Salt Spring Island, British Columbia.

Notes The common name comes from the Greek rhodon, meaning "rose," referring to the mineral's distinctive color. Rhodonite is often used as a gemstone (see p. 4) because of its pale to hot pink coloring. It is a brightly colored mineral that nearly always has a crust of black manganese oxide hiding its true colors. Saltspring Island and Vancouver Island, British Columbia, are both known for their rhodonite deposits.

Pink feldspar is similar in appearance, but it does not have narrow veins of dense black manganese oxide present, as rhodonite does.

Serpentine

Composition $Mg_3Si_2O_5(OH)_4$ • Magnesium iron silicate hydroxide (Lizardite)
Classification Minerals – Silicates. **Color** Olive-green to blackish, yellowish green, brown, yellow or (rarely) white. **Luster** Silky, waxy to greasy. **Transparency** Translucent to opaque. **Streak** White. **Cleavage** Perfect. **Fracture** Conchoidal (antigorite). **Crystals** Monoclinic; variable depending upon variety. **Formation** Formed by hydrothermal metamorphic processes. **Hardness** 2.5-4. **Cleavage** Varies.

British Columbia.

Notes The serpentine mineral group consists of basic magnesium silicates, including antigorite, chrysotile (see p. 48) and lizardite. These minerals all have similar formulae.

The serpentine minerals are not always easily identifiable. Antigorite is lamellar (layered) and normally occurs in compact masses in a shade of dark green. Chrysotile is usually in fibrous yellowish or green aggregates (asbestos). Lizardite displays its minute scales or compact whitish aggregates.

Stilbite

Composition $NaCa_2Al_5Si_{13}O_{36}.14H_2O$ • Hvdrated sodium calcium aluminum silicate **Classification** Minerals – Silicates. **Color** White, pink, red, brown or orange. **Luster** Vitreous. **Transparency** Transparent to translucent. **Streak** White. **Cleavage** Perfect. **Fracture** Uneven. **Crystals** Monoclinic; bladed aggregates with numerous twinned crystals, pinched in the middle to resemble wheat sheaves; also radial. **Formation** Forms in cavities of some plutonic or volcanic rocks. **Hardness** 3.5-4. **Cleavage** Perfect.

Roseburg, Oregon.

Notes Stilbite is a member of the zeolites group of minerals. Zeolites are used in industry for their performance as filters and chemical sponges. They are also very popular with collectors for their beauty.

Stilbite's extraordinary crystals often grow into amazing shapes that resemble wheat sheaves or stacks of bow ties.

Talc

Composition $Mg_3Si_4O_{10}(OH)_2$ • Magnesium silicate hydroxide **Classification** Minerals – Silicates. **Color** White to dark green, gray or brownish. **Luster** Pearly or greasy. **Transparency** Translucent. **Streak** White. **Cleavage** Perfect. **Fracture** Uneven. **Crystals** Monoclinic; crystals rare; usually foliated, granular. **Formation** Characteristically found in metamorphic rocks that originated from dolomite rock and peridorite. **Hardness** 1. **Cleavage** Perfect.

Mahood Lake, British Columbia.

Notes This mineral is extremely soft—there are not many minerals with a hardness of 1. It usually forms grains of flaky masses and on rare occasions a crystal.

The uses of talc vary, from the best-known use as talcum powder, to other cosmetic uses, to its use as a lubricant and as insulation for electrical equipment.

Thomsonite

Thomsonite with Mesolite, Bear Creek quarry, Oregon.

Composition $NaCa_2Al_5Si_5O_{20}.6H_2O$ • Hydrated sodium calcium aluminum silicate **Classification** Minerals – Silicates. **Color** Colorless, white, yellowish, pink or greenish. **Luster** Vitreous to pearly. **Transparency** Transparent to translucent. **Streak** Colorless. **Cleavage** Perfect. **Fracture** Uneven. **Crystals** Orthorhombic; normally lamellar or radiating aggregates; also prismatic crystals. **Formation** Found in basalt cavities. **Hardness** 3.5-4. **Cleavage** Perfect.

Notes Thomsonite is a member of the zeolites group of silicate minerals. Distinct crystals of thomsonite are rare, as it usually forms radiating balls or globular aggregates. Its color changes to brown when impurities are present.

This mineral gelatinizes with hydrochloric acid. To test your specimen for thomsonite, carefully place a drop of the acid on a small portion of the mineral.

In western North America, thomsonite it has been found at Table Mountain, Colorado, and several locations in Oregon.

Thomsonite with Calcite, Yellow Lake, British Columbia.

Titanite (Sphene)

Composition $CaTiSiO_5$ • Calcium titanium silicate **Classification** Minerals – Silicates. **Color** Brown to black, gray, yellow or green. **Luster** Vitreous. **Transparency** Transparent to nearly opaque. **Streak** White. **Crystals** Monoclinic; usually tabular crystals; also lamellar, compact. **Fracture** Conchoidal. **Formation** Titanite forms in many plutonic rocks and in some metamorphic rocks. It also occurs as small crystals in granite, granodiorite and diorite of plutonic rocks. **Hardness** 5-5.5. **Cleavage** Distinct.

Whitehall, Montana.

Notes Titanite, named for its titanium content, varies considerably from transparent to nearly opaque. Several brown translucent crystals reaching 1″ (2.5 cm) across form with adularia and black tourmaline near Whitehall, Jefferson Co., Montana. Titanite is also known as sphene, and it is classified as a gemstone.

Topaz

Composition $Al_2SiO_4(OH,F)_2$ • Aluminium fluohydroxisilicate **Classification** Minerals – Silicates. **Color** Colorless to pale yellow or blue, greenish, pink. **Luster** Vitreous. **Transparency** Transparent to translucent. **Streak** White. **Fracture** Conchoidal. **Crystals** Orthorhombic; normally in stubby to medium-long prismatic crystals; also granular, massive. **Formation** Normally found in igneous rocks and high-temperature veins. **Hardness** 8. **Cleavage** Perfect.

Notes Topaz is so hard that it is sometimes found deposited in river gravels, washed there from a weathered parent rock. Water-worn pebbles of colorless topaz are sometimes mistaken for diamonds. Topaz is the birthstone for November.

San Diego County, California.

Tremolite

British Columbia.

Composition $Ca_2(Mg,Fe)_5Si_8O_{22}(OH)_2$ • Calcium magnesium silicate hydroxide **Classification** Minerals – Silicates. **Color** Colorless, or white, gray, green, pink or brown. **Luster** Vitreous or silky. **Transparency** Transparent to translucent. **Streak** White. **Fracture** Uneven. **Crystals** Monoclinic; usually fibrous, radiating aggregates; also bladed, compact. **Formation** Forms in contact with metamorphosed dolomites and in serpentinites. **Hardness** 5-6. **Cleavage** Perfect.

Notes Tremolite crystals are often long bladed or prismatic, up to 3" (7.5 cm) in length. They may have a fibrous or massive habit. The fibrous form is referred to as asbestos, along with other similar fibrous amphiboles (a group of closely related, rock-forming silicate minerals). Asbestos was formerly used in industry as an insulator.

Tremolite also combines with actinolite (see p. 61) to produce nephrite, one type of jade.

Zoisite

Greenhorn Mountains, Montana.

Composition $Ca_2Al_3(SiO_4)_3(OH)$ • Calcium aluminum silicate **Classification** Minerals – Silicates. **Color** Gray, yellowish brown, greenish or pink. **Luster** Vitreous. **Transparency** Transparent to translucent. **Streak** White. **Cleavage** Perfect. **Fracture** Uneven to conchoidal. **Crystals** Orthorhombic; normally in long, prismatic crystals with deep striations; also columnar, bladed, fibrous, compact. **Formation** Forms in metamorphosed sediments and granites. **Hardness** 6.5-7. **Cleavage** Good.

Notes Zoisite is a mineral that forms in high-temperature and high-pressure conditions in metamorphic rocks. The manganese-bearing pink variety of zoisite, thulite, is an ornamental stone that displays fluorescence.

MINERALOIDS

Mineraloids are mineral-like solids that occur naturally on Earth but do not conform with the basic properties of minerals. They do not fit into any of the chemical families since they rarely form crystals and are typically organic in origin. Mineraloids are formed from fossilized or compacted living matter.

Amber

Amber in coal, Blakeburn, British Columbia.

Composition Approximately $C_{10}H_{16}O$ • Succinic acid **Classification** Mineraloids. **Color** Varies from gold through amber to brown; occasionally green, violet or black. **Luster** Resinous or subvitreous. **Transparency** Transparent to translucent. **Streak** White. **Fracture** Conchoidal. **Crystals** Usually amorphous (no crystal structure). **Formation** Forms from the resin of coniferous trees; also found in sedimentary rocks. **Hardness** 2.5. **Cleavage** None.
Notes Amber is the fossil resin of extinct coniferous trees. It may contain various fossilized insects, very small vertebrates and plants that were trapped in the original sticky resin. This very soft mineraloid is often found in coal deposits.

Aragonite

Ammolite fossil, Alberta.

Composition $CaCO_3$ • Calcium carbonate **Classification** Mineraloids. **Color** White, colorless, gray, green, blue, violet, reddish, yellowish and brown. **Luster** Vitreous, resinous. **Transparency** Translucent to transparent; fluorescent. **Streak** White. **Fracture** Subconchoidal. **Crystals** Orthorhombic, usually prismatic; twinned crystals common. **Formation** In Europe, aragonite forms around hot springs, in mineral veins and as stalactites in caves. In North America it is the mineral that causes iridescence on the shells of molluscs. **Hardness** 3.5-4. **Cleavage** Good.
Notes Aragonite occurs in evaporate deposits of chemical sedimentary rocks—a source that is rare in North America. Aragonite also produces the mother of pearl on ammolite and nacreous pearls. Ammolite is the organic gemstone that is formed from fossilized extinct molluscs called ammonites (squid-like marine animals). In 1981, ammolite was officially recognized as an organic gemstone.

Ammolite is only found in one area in the world—along the eastern Rocky Mountains of Alberta and Montana. It is a rare mineral that is present in only 5% of the ammonite shells that are mined there.

Opal

Opal, thunderegg, Opal Butte, Oregon.

Composition SiO_2,nH_2O • Silicon dioxide **Classification** Mineraloids. **Color** Precious opal is milky white or black, with a splendid display of iridescence highlighting red, blue, yellow and other colors. Other varieties occur in a wide array of colors but do not display iridescence. **Luster** Pearly to resinous or greasy. **Transparency** Transparent to translucent. **Streak** White. **Fracture** Conchoidal, conspicuous when compact. **Crystals** Amorphous (no crystal structure). **Formation** Opal is formed from silica solutions resulting from the deposition of precipitation in openings in volcanic or sedimentary rocks. Opal is also commonly found as the replacing medium (replacing the fibers) in petrified wood. **Hardness** 5.5-6.5. **Cleavage** Uneven
Notes Opal is sometimes classified as a silicate variety of mineral, but it lacks crystals. It is made of silicon dioxide, like quartz, but with an average of 10% water (ranging from 5 to 20%). As with all precious stones, the quality of opal is measured in carats.

Solid Opal Semi-crystal, Opal Resources Canada Inc., Vernon, British Columbia.

Opals should never be subjected to large changes in temperature because it may cause them to crack. Commercial precious opal is normally capped with a clear covering of quartz to protect the stone. Opal is the birthstone for October.

Potch opal, a type of common opal, is dull yellow, red or black. It is mined for use as an abrasive and filler.

VARIETIES OF OPAL

Common opal is milky, black, yellow, brick red or green, with no iridescence. It is widespread and abundant.

Fire opal is clear orange or reddish, and often has a rainbow display of colors.

Precious opal, found in British Columbia and Idaho, shows a striking rainbow of colors. The electron microscope reveals the structure of precious opal: regularly spaced tiny spheres, which are responsible for the light interference that produces the color display. This quality of precious opal is treasured worldwide.

Hyalite opal is a colorless, light yellow or transparent blue variety of opal that lacks a play of color. Hyalite fluoresces green when illuminated with ultraviolet light.

Hyalite Opal, Riverside County, California, fluorescent display under ultraviolet light.

Hyalite Opal, Riverside County, California, display in daylight.

Boulder opal forms in vertical and horizontal cracks of boulders, or as coatings on boulders. This opal is a very desirable one, known as a designer's delight for its ironstone inclusions in the foreground and all sorts of amazing shapes.

Wood opal replaces the cells in fossilized wood without changing the texture or detail.

Rough Opal in matrix, British Columbia.

Boulder Opal, polished & rough cut.

ROCKS

Basalt Columns, Sheepeater Cliff, Yellowstone National Park, Wyoming.

IGNEOUS ROCKS

Rock Types
Rocks are classified into three major groups: igneous, sedimentary and metamorphic.

IGNEOUS ROCKS
Igneous rocks are volcanic in nature, forming from cooling magma (molten rock). They are divided into two groups.

Extrusive Igneous Rocks
When lavas flow out onto the surface of the Earth, they are called extrusives, or extrusive igneous rocks. When cooling occurs on the Earth's surface, the material cools very quickly, forming volcanic igneous rocks. The crystals produced are very small, because they do not have sufficient time to grow large. A good example is basalt.

Intrusive Igneous Rocks
When magma flows deep into fissures, it cools slowly and the crystals can grow much larger. The result is plutonic or intrusive igneous rocks. Their composition depends upon the makeup of the molten rock.

The chemical makeup of each igneous rock depends largely upon the elements present in the magma at the time of the eruption.

Andesite

Classification Igneous (Extrusive). **Color** Brown, greenish or dark gray. **Texture** Fine-grained. **Origin** This rock forms as lava flows from andesitic volcanoes and as shallow dike intrusions. **Grain Size** Fine.

Colorado.

Notes Andesite contains 55 to 65% silica content. It occurs primarily as lava flows following an eruption, and also when it is injected into dike intrusions and domes. It is very common, in abundance second only to the volcanic rock basalt. It is composed primarily of plagioclase with other minerals, including hornblende, pyroxene and biotite.

Anorthosite

Wyoming.

Classification Igneous (Intrusive). **Color** White or light gray. **Texture** Granular, and somewhat similar to granite. Banded sections are separated by grain size and textre. **Origin** Forms in intrusive environments, in stocks and dikes. **Grain Size** Coarse.
Notes This common rock is a type of diorite that is composed almost entirely of feldspar and virtually no quartz. This rock is basic in nature (it contains between 45 and 55% total silica). In the west, it is present in the Stillwater Complex of Montana. It has also been discovered to form part of the moon's surface.

Diorite

British Columbia.

Classification Igneous (Plutonic). **Color** Normally gray to dark gray, occasionally greenish to brownish gray. **Texture** Appears similar in texture to granite, but normally darker. **Origin** Normally forms major granitic masses but may also be encountered as independent intrusions, such as dikes. **Grain Size** Coarse.
Notes Diorite contains a mixture of feldspar, pyroxene, hornblende and a small percentage of quartz (less than 5%). This rock is a variety of gabbro (see p. 71). It exhibits an intermediate chemical composition between an acid and base. Deposits of diorite have been found in Kern and Tuolumne Counties, California.

Basalt

Basalt pebble, Nootka Island, British Columbia.

Classification Igneous (Extrusive). **Color** Dark. **Texture** Fine, with crystalline structures too small to be distinguished with the unaided eye. **Origin** Forms when magma erupts on the Earth's surface with rapid cooling and hardening of the lava flows. **Grain Size** Very fine. **Notes** Basalt is one of the most widespread volcanic igneous rocks on Earth. It is also abundant on the moon. Most of the moon's surface is covered with basaltic lava flows and flood basalts.

Basalt has a wide variety of uses, including aggregate crush for road construction, concrete aggregate, asphalt pavement aggregate, railroad ballast and many others.

Basalt, with large feldspar phenocrysts formed in magma chamber before being expelled.

Granite

Classification Igneous (Intrusive). **Color** Red, pink, gray or white, dark mineral grains visible throughout. **Texture** Coarse, with a rough texture. **Origin** Granite forms from the slow crystallization of magma below the Earth's surface. **Grain Size** Coarse.

Notes Granite exhibits an acidic chemical composition. It is composed mainly of quartz and feldspar with minor amounts of micas, amphiboles and other minerals.

Granite is cut to various sizes and used in interior and exterior construction. It resists abrasion and is strong enough to bear significant weight. Granite also withstands weathering, and it accepts a brilliant polish. These characteristics make it a very desirable construction material. Granite is quarried in western North

White Granite, British Columbia.

White Granite pebble, British Columbia.

Pink Granite, Georgetown, Colorado.

Granodiorite

Classification Igneous (Intrusive). **Color** Light to dark gray, less commonly pink. **Texture** Normally massive in structure and frequently with a flow orientation. **Origin** Formed by the slow cooling and crystallization of magma deep in the Earth's crust. **Grain Size** Medium to coarse.

Notes Granodiorite is likely the most common rock of the granite "family." It is very similar to quartz, having the same light-colored essential minerals but in different proportions. The quartz content is the same as granite (10-30%), but the plagioclase feldspars (oligoclase-andesine) are more abundant than the alkali feldspars (orthoclase, minocline, and perchite).

Granodiorite exhibits an intermediate chemical composition between an acid and base.

Canada.

Gabbro

Classification Igneous (Plutonic). **Color** Black or gray to very dark green, and (rarely) reddish. Some specimens appear mottled in the hand. **Texture** Coarse grain has an even texture and is often speckled with minerals. **Origin** Forms in plutonic intrusions, which are normally layered. **Grain Size** Coarse.

Wyoming.

Notes The mineral makeup of gabbro is very similar to that of basalt. It is composed mainly of feldspar with smaller amounts of pyroxene and olivine. Gabbro often contains noteworthy amounts of some relatively rare metals, including titanium, nickel, chromium and platinum. In western North America, prominent mineral deposits of gabbro have been found in Nevada Co., California.

Obsidian

Oregon.

Classification Igneous (Extrusive). **Color** Normally black but can be gray, reddish brown, mahogany or dark green. **Texture** Smooth, lustrous and glassy. **Origin** Formed by the very rapid cooling of viscous acid lava. **Grain Size** None.

Notes Obsidian is a silica-rich volcanic rock. It forms when a magma containing a high percentage of silica flows to the Earth's surface. Obsidian does not contain any crystals, because it has cooled too quickly for any to group into the regular arrangements of minerals. Obsidian has been used for making cutting tools since Palaeolithic times.

Snowflake Obsidian tumble polished, Utah.

With its hardness of approximately 5.5, obsidian is relatively easy to carve. Artists have used it to carve small sculptures and make masks for thousands of years. Today, rock enthusiasts sometimes tumble-polish local specimens. One distinctive type is snowflake obsidian which displays microcrystalline patches of white color.

Pegmatite

Classification Igneous (Plutonic)
Color Pale pink or whitish. **Texture** Varied, with much zonation and many vugs (open cavities). **Origin** Forms deep within the Earth, often in dikes and veins. Often concentrated at the margins of granite intrusions. **Grain Size** Very coarse-grained, with large crystals.

Notes Pegmatites are among the most fascinating types of rock in the world. No other rock formation contains such an array of spectacular crystals. All of the world's largest natural crystals have been found in pegmatites.

One famous pegmatites location in western North America is the noteworthy Himalaya Mine, Pala Co., California. Another site in South Dakota produced a gigantic spodumene crystal an amazing 42′ (13 m) long.

Colorado.

Porphyry

Classification Igneous (Extrusive), (Intrusive) or (Plutonic).
Color Variable. **Texture** A mixture of fine and coarse grains. **Origin** Volcanic. **Grain Size** Phenocrysts (large crystals) are dispersed throughout a fine-grained rock.
Notes Porphyry is a general term for igneous rock that is identified by its texture rather than its minerals, which vary considerably. Larger crystals, called phenocrysts, are embedded in a fine-grained rock. The large crystals are commonly feldspar. Many types of porphyry may be encountered. A few of the common types are flowerstone, snowflake and Chinese writing.

Flowerstone is one type of porphyry. These gemstones originate from Texada Island and Vancouver Island, BC.

Porphry pebble, British Columbia.

Flowerstone, Vancouver Island, British Columbia.

Pumice

Nevada.

Classification Igneous (Extrusive). **Color** Normally light gray, but can be yellowish or red. **Texture** Vesicular (pitted with many cavities inside and out). It is very light and it floats on water. **Origin** Associated with rhyolitic volcanic eruptions. **Grain Size** Fine.

Notes Pumice is an extremely porous, brittle type of rhyolite (see below). This frothy volcanic rock forms from lava that is full of gas. The lava is ejected and shot into the air during an eruption. As it hurtles through the air it cools rapidly, solidifying, and the gases escape, leaving the rock full of cavities. Giant pieces of pumice have been seen floating on the ocean after major eruptions. The largest of these lava blocks are big enough to carry small mammals to a new location. Today, pumice is ground up and used in soaps, abrasive cleansers and polishes.

Rhyolite

Thomas Mountains, Utah.

Classification Igneous (Extrusive). **Color** Light. **Texture** Phenocrysts (large crystals) embedded in a fine-grained volcanic rock. Rhyolite may also have vesicles and amygdales present. **Origin** From volcanoes with explosive violence, and the cooling of viscous lava. **Grain Size** Fine.

Notes Rhyolite is a fine-grained acidic rock that is closely related to granite and typically contains the minerals quartz and feldspar. The structure of rhyolite is quite variable, from pimply (like pumice) to fluidal, perlitic (containing water), spherulitic (small, rounded bodies) and massive. The vesicular structures of rhyolite are distinctive when present.

SEDIMENTARY ROCKS

Sedimentary rocks are formed by the erosion, transportation and deposition of existing rocks. These rocks are relatively easy to recognize because they normally have a layered structure. They are divided into three groups.

Detrital (Clastic) Rock
Rock formed by the accumulation of fragments of pre-existing rocks and minerals, deposited by water, wind, ice and other mechanical agents. Examples include sandstone and conglomerate.

Chemical Rock
Rock formed by chemical sediments from minerals precipitated directly from a saline or freshwater solution. Examples include rock salt and gypsum rock.

Biogenic (Organic) Rock
Organic sedimentary rock formed directly or indirectly through the action of plants or animals. Examples include coal and coral limestone.

Arkose

Classification Sedimentary (Clastic). **Color** Varies from gray to pink or reddish. **Texture** Coarse grain size, not well sorted and having poor stratification. **Origin** Forms in continental deposits and aquatic environments, both marine and freshwater. **Grain Size** Medium.

Colorado.

Notes Arkose is made up primarily of quartz, but feldspar can contribute as much as a third of this rock. In terrestrial environments, most arkoses are derived from granite disintegration.

In the Californian Coast Ranges, most arkose rocks were formed between the Jurassic and the present day.

Bituminous Coal

Classification Sedimentary (Organic). **Color** Iron-black to velvet-black. **Texture** Hard and brittle with a dull luster. **Origin** Coal originates from the accumulation of plant debris, usually in a swamp environment where it decays. Due to a lack of oxygen, this matter persists. After a thick layer of debris is formed, it is buried by sediments such as mud or sand. Approximately ten feet of plant debris is needed to be compacted into just one foot of coal. **Grain Size** Medium, fine.

Notes Coal is formed by the long-term process of plant decay with rising pressures and temperatures. Bituminous coal (soft coal) is the most abundant fossil fuel found on Earth. It generates the highest heat value but burns with a yellow flame that gives off smoke and a distinctive odor. The carbon content of bituminous coal ranges from 50 to 65%. Anthracite (hard coal) is formed from bituminous coal that is subjected to greater pressures and temperatures in a metamorphic environment. Anthracite coal has a fixed carbon content of 85 to 95%.

Alberta.

Breccia

Colorado.

Classification Sedimentary (Clastic). **Color** Breccia can be any color. **Texture** Very angular fragments of rock that have not been smoothed by water. **Origin** The large, angular fragments typically accumulate where mechanical weathering is active. **Grain Size** Very coarse.

Notes Breccia fragments are angular, and can be from igneous, metamorphic or sedimentary rocks. The fragments are bound together in a fine- to medium-grained matrix of sand. Breccia is often found at the bottoms of cliffs.

Gypsum Rock

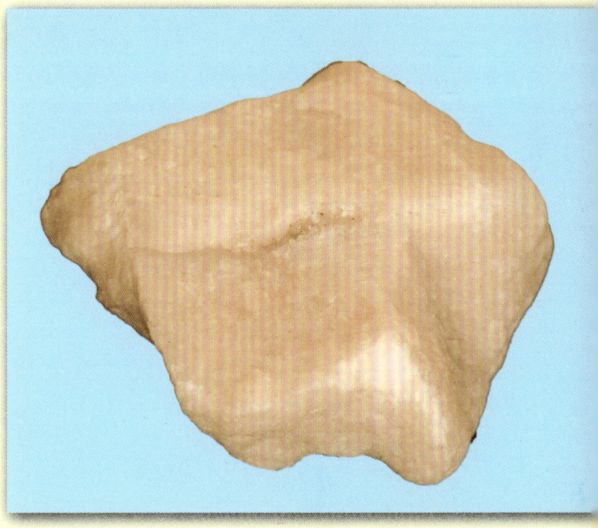
Colorado.

Classification Sedimentary (Chemical). **Color** Usually white to light gray; may be yellow or red as a result of iron oxide; may be dark gray from clay or organic impurities. **Texture** Fine to medium grain. **Origin** Gypsum rock originates when saline lakes or sea waters evaporate, in an arid or semi-arid climate. **Grain Size** Fine to medium.

Notes Gypsum is associated with anhydrite, rock salt and other marine salts; and calcite, dolomite, clays and limonite. It is commercially very important for the production of plaster of Paris and gypsum wallboard (drywall) for construction.

Gypsum rock is mined in British Columbia, Alberta, California, Utah and Montana.

Conglomerate

Colorado.

Classification Sedimentary (Clastic). **Color** Varies greatly due to the large size and varied colors of the minerals, rock fragments and a finer-grained matrix present. **Texture** Large grains are obviously rounded. **Origin** Forms in beaches, river systems and other environments with sufficient energy to move large fragments of material. **Grain Size** Very large.

Notes Conglomerate rocks can be useful in determining the presence of desirable minerals in the area. For example, kimberlite is a mineral that hosts diamonds. so if it is present in a conglomerate, then it is to be found upstream—somewhere.

Conglomerate rocks are not commercially utilized to a great extent.

Dolomite Rock

Colorado.

Classification Sedimentary (Chemical). **Color** Similar to limestone, but more often light gray to tan. **Texture** Similar to limestone, often compact and earthy. **Origin** Forms in various marine environments. Scientists believe that most dolomites are of secondary origin, replacing the original limestones. **Grain Size** Medium, fine.

Notes Dolomite is the name of a rock and also a mineral. The rock contains a high proportion of the mineral dolomite, from which it gets its name. Dolomite rock is rarely a mineral by itself. It is readily subject to the effects of metamorphism.

Limestone

Colorado.

Classification Sedimentary (Organic). **Color** Normally white to dark gray or black, very occasionally red. Darker colors are caused by organic materials, and yellow and brown colors result from iron oxides. **Texture** Appears as a dense rock with an even, granular texture. **Origin** Varies widely depending on environment. **Grain Size** Medium fine.

Notes Limestone rock consists of a minimum of 50% calcite (or aragonite). The remainder consists of a variety of impurities, which may include grains of dolomite, quartz, clay and others. Limestone can also be reprecipitated in fractures, caves and similar locations. Stalactites, stalagmites and crystalline "sculptures" at hot springs and geysers are a few examples.

Commercially, limestone is used primarily in construction material, as crushed stone for roads and railroad ballast, and as an aggregate in concrete. Limestone is also fired in a kiln with crushed shale to make cement.

Fossiliferous Limestone

Classification Sedimentary (Organic). **Color** Gray and brown. **Texture** A rough rock, medium to fine grain. **Origin** Normally marine in origin, but a few of these limestones form in freshwater environments. **Grain Size** Medium, fine.

Notes Fossiliferous limestone contains a high proportion of fossil shells and can contain a great variety of both brachiopod and bivalve shells. The actual rock matrix is usually cemented by calcite. If a brownish coloring is visible, it may be due to the presence of heavy minerals and iron oxides. Shelly limestone is another common name for fossiliferous limestone.

Alberta.

Shale

Alberta.

Classification Sedimentary (Clastic). **Color** A variety of colors. Black and gray shales are the most common, but red, brown, yellow and green shales may also be encountered. **Texture** Dense, with particles the size of silt and clay. Shale is compact, hard and thin-bedded. **Origin** Deposited from sediments in lake and ocean bottoms. **Grain Size** Very fine.

Notes Shale is a common rock found in many areas of North America. It is used in the manufacture of cement, and in bricks, pottery and other ceramic products, including the terra-cotta pots you may have in your home.

Black shales contain organic material that can break down to form natural gas or oil, so shale formations are important sources of these resources.

Sandstone

Classification Sedimentary (Clastic). **Color** Varies greatly from white to gray, yellowish, or dark red. **Texture** Normally gritty to the touch. **Origin** The most common type of sandstone is formed by accumulations in marine waters, or as windblown deposits in arid continental areas. A finer-grained matrix binds these grains of sand together over millions of years. **Grain Size** Medium.

Notes The color of sandstone depends upon the cement (medium) that glues the sand grains together, as well as the overall color of the mineral grains.

Sandstone is used primarily in construction as it is easy to work with. It can also be quarried into large, uniform-size blocks, often called freestones.

California.

Sandstone pebble, British Columbia.

Fossiliferous Sandstone, Beverly Beach, Oregon.

Siltstone

Utah.

Classification Sedimentary (Clastic). **Color** Black to gray or sandy brown. **Texture** Fine-grained. The individual rock fragments and mineral grains are too small to be visible to the naked eye. **Origin** Siltstone is produced through the compaction of silt sediments that have accumulated in a marine or freshwater environment. **Grain Size** Fine.
Notes Siltstone contains twice as much silt as clay. The less clay that is present in siltstone, the more soft and crumbly it is. Various embedded fossils may be found, including invertebrates, vertebrates and plants.

Mudstone

Alberta.

Classification Sedimentary (Clastic). **Color** Black, white, grey, brown, red, green or blue. **Texture** Dense, with particles of silt size (microscopic) and clay size (submicroscopic). Mudstone is a hardened rock but may break apart when wet. **Crystals** Extremely small, "clay sized." **Origin** From sediments deposited in the quiet environments of lake and ocean bottoms. **Grain Size** Very fine.
Notes Mudstone is a general term for a number of rocks that contain approximately 33% silt grains and 66% ultra-fine clay grains. It is actually hardened mud, and like most mud it often includes plant and animal matter. This organic matter tends to turn the stone black.

Mudstone is generally too soft to be of much commercial use.

METAMORPHIC ROCKS

Metamorphic rocks are formed by the physical and chemical changes—mainly heat and pressure—of existing igneous and sedimentary rocks. Three groups of metamorphic rocks can be identified.

Contact Metamorphic Rocks
Contact metamorphism is caused by direct heat that usually creates crystalline rock (hornfels).

Regional Metamorphic Rocks
Regional metamorphism is a result of both heat and pressure, which cause foliation (parallel layers) or cleavage in rocks where the minerals have been aligned.

Cataclastic Metamorphic Rocks
Cataclastic metamorphism occurs where rocks are being faulted and sheared. This is not a common type of rock.

Amphibolite

Classification Metamorphic (Regional). **Color** Dark green, often with white and yellow speckles or patches. **Texture** Fibrous minerals oriented in the same direction give this rock its texture. **Origin** A metamorphic rock that forms through recrystallization under high viscosity and directed pressure. **Grain Size** Coarse.
Notes Amphibolite is mainly composed of the minerals amphibole and plagioclase. Other minerals may also be present, including feldspar, pyroxene, chlorite, epidote and garnet. Amphibolite is often linked to copper deposits.

Colorado.

Eclogite

Classification Metamorphic (Regional) **Color** Varies from green to red, and often speckled. **Texture** A very fine- to coarse-grained rock, often with compositional zonation. **Origin** Formed under the highest temperature and pressure conditions, at great depths in the Earth's crust. **Grain Size** Very fine to coarse. **Notes** Eclogite is found in the Precambrian Lewisian rocks in the Jurassic Franciscan Formation of California. It has no commercial value.

Russia River, California.

Gneiss

Classification Metamorphic (Regional) **Color** Normally gray, tan or pink. **Texture** Granular (mixed), medium- to coarse-grained, crystalline, with a somewhat parallel mineral orientation. **Origin** Gneisses originate from a sedimentary rock such as sandstone or shale, or the igneous rock granite. The segregation of minerals into bands is the result of very high temperatures (450° to 700°C) and corresponding high pressures. **Grain Size** Coarse. **Notes** Gneiss is composed of the minerals feldspar and quartz. Gneiss is subjected to more heat and pressure than a schist, and it has distinct banding from alternating layers of different minerals. Gneiss is coarser than schist. Some examples show intense folding and deformation, which occurs as a result of its molten state when deep within the Earth's crust.
Some gneissic rocks are used as building stones and paving stones.

Georgetown, Colorado.

Hornfels

Classification Metamorphic (Local/Contact). **Color** Dark gray to black, also pink, brown, violet and green. **Texture** Varies greatly from fine to medium granular. **Origin** Formed from a channel of rising lava and saturated with volcanic gases and solutions, then thrown to the Earth's surface. **Grain Size** Fine to medium.

Notes Hornfels is a nonfoliated (unbanded) metamorphic rock without a specific composition. It is a rock that was "baked" while near a heat source such as a magma chamber, sill or dike. Examples can be found at the Sierra Nevada of California and the Mackay region of Idaho.

Wyoming.

Phyllite

South Dakota.

Classification Metamorphic (Regional) **Color** Silvery gray to lead gray; occasionally green (from chlorite), or red, purple, brown or yellow (from iron oxides) with a dull micaceous luster. **Texture** This foliated rock consists of fine to medium grain size; small, distinct crystals may be present.

Origin Forms from sediments, during low temperature and low to moderate pressure. **Grain Size** Fine.

Notes Phyllite is very similar to slate, except that the metamorphism is more intense and the new mica crystals grow larger to give a distinct micaceous parting direction to the rock.

Phyllite is a metamorphic rock that is made up primarily of very fine-grained mica. Its surface is typically lustrous and occasionally wrinkled. It is intermediate, between slate and schist. Phyllite is easily scratched with a knife blade. It is quarried for roofing slate and flagstone.

Marble

Classification Metamorphic (Dynamic, Regional or Contact). **Color** Varies greatly from very white to variously patched or streaked with green, gray, brown and red. **Texture** Varies greatly from fine to coarse grain. **Origin** A metamorphic rock that is produced from the metamorphism of limestone. **Grain Size** Fine- to medium-grained. **Notes** Marble is composed primarily of calcium carbonate. With an estimated hardness of 3, marble can easily be scratched with a knife blade.

British Columbia.

Marble is a very important material in construction, both as rough stone and as polished slabs. It is also commonly used as a raw material for sculpture (statuary marbles). Noteworthy sources have been found in California, in Fresno, Inyo, Kern, Madera and Riverside Counties.

Quartzite

Classification Metamorphic (Regional). **Color** White to black, also buff, brown and red. **Texture** Even consistency with primarily rounded granules of sand, held together by cementation by silica, carbonates, clay or iron oxides, or compaction in the presence of clay. **Origin** Formed from sediments accumulated in a wide variety of environments, including beaches, deltas, floodplains and deserts. **Grain Size** Medium.

Wyoming.

Notes Quartzite is made from sandstone that has been metamorphosed. It looks like sandstone, but it is much harder because it forms from sandstone that has come into contact with very deep magmas.

The best way to tell the difference between the two is to break the rock. Sandstone shatters into many individual grains of sand, while quartzite breaks across the grains of sand.

Quartzite is used primarily in the construction industry for flooring, facing and similar uses.

Schist

Classification Metamorphic (Regional). **Color** Silvery white to gray, also yellow to brown tones. Color varies with the type of mineral or minerals present. **Texture** Varies from medium- to coarse-grained, with prominent parallel flaky or tabular mineral orientation. **Origin** Schists are formed from shale or mud, at higher temperatures than slate. **Grain Size** Medium.

Schist & green Actinolite needles intergrown, Washington.

Notes Schists are usually named for the main mineral from which they are formed. Garnet schist is rich in micas, biotite and muscovite. Well-shaped crystals of garnet are about 0.2" (5 mm) in diameter, and have grown in the rock during slow pressure and temperature changes. The garnet is usually the reddish variety due to the presence of almandine.

Schists have been found in Alaska (Wrangell district), California (Inyo, Madera, Marin, Mendocino, San Benito and Sonoma Counties) and Washington (N. Cascades).

Serpentinite

Ocean-picture rock, British Columbia.

Classification Metamorphic (Regional). **Color** Varies from green to yellow, to reddish brown, to black. **Texture** Dull to waxy; often somewhat greasy and very smooth to the touch. **Origin** Derived from the metamorphism of a wide variety of minerals in a water-rich environment. **Grain Size** Fine.
Notes Serpentinite, a widespread rock, contains metamorphic crystals that can only be seen with the aid of a microscope. It has been found in many places, including in California, Oregon and Montana. Ocean-picture rock is a local name for a brownish, brecciated serpentinite with streaks of pale blue to white quartz. This variety of rock is found near Midway, British Columbia.

Slate

Classification Metamorphic (Regional). **Color** Normally light to dark gray, but also in shades of green, red, purple, brown or yellow. **Texture** A fine-grained rock with a characteristic perfect, slaty cleavage produced by the alignment of flaky minerals such as mica. This makes it easy to cut into parallel slabs. **Origin** Forms when fine-grained sediments such as mudstones or shales are metamorphosized at low temperatures and low pressures. **Grain Size** Fine.
Notes Slate is shale or mudstone that has been metamorphosized (hardened). Tiny mica flakes have grown along new cleavage surfaces to give the hardened shale a luster. These flakes are arranged so that their flat sides lie across the direction of the pressure. The result is that the cleavage of the slate follows the new mica plate direction, at right angles to the squeezing.

Slate is used in construction for interior flooring, exterior paving, dimension stone and decorative aggregates.

Rock & Lapidary Clubs

CANADA

Gem & Mineral Federation of Canada
http://www.gmfc.ca/gmfcmembers.htm

Canadian Geological Clubs & Associations
http://www.canucklinks.com/gem/geoclubsnosite.htm

British Columbia

Club Directory of the British Columbia Lapidary Society
http://www.lapidary.bc.ca/directory.html

Alberta

Alberta Federation of Rock Clubs
http://www.afrc.ca/

Saskatchewan

Prairie Rock & Gem Society
http://prairierockandgem.info/

UNITED STATES

Mineral, Rock, and Gem Clubs & Related Organizations – USA (The-Vug.com)
http://www.the-vug.com/vug/vugclubs.html

American Federation of Mineralogical Societies (AMFS)
http://www.amfed.org/

Friends of Mineralogy
http://www.friendsofmineralogy.org/

Washington

Washington State Gem and Mineral Clubs
http://www.dnr.wa.gov/RecreationEducation/Topics/HarvestingCollecting/Pages/gem_clubs.aspx

Oregon

Oregon Gem and Mineral Clubs
http://www.rockhounds.com/rockshop/clubs/oregon.shtml

Alaska

Chugach Gem and Mineral Society
http://chugachgms.org/

California

California Federation of Mineralogical Societies
http://www.cfmsinc.org/

Rocky Mountain Region

Rocky Mountain Federation of Mineralogical Societies
http://www.rmfms.org/

Arizona

Prescott Gem & Mineral Club – Prescott Valley, Arizona (one of many Arizona Clubs)
http://www.prescottgemmineral.org/

Glossary

accessory mineral: a mineral constituent of a rock that occurs in small quantities and is disregarded in the definition and chemistry.
acicular crystals: very elongated crystals with a needle shape.
acidic rock (acid rock): a type of igneous rock (e.g., granite) that is made up of predominantly light-colored minerals and more than 66% silica.
aggregate: an assemblage of crystals of one or more species.
amorphous: mineral-like material with no crystal structure (e.g., opal).
amphiboles: a group of closely related, dark-colored, rock-forming silicate minerals.
amygdules: deposits that form when gas bubbles or vesicles are filled with a secondary mineral.
anhedral crystal: a poorly formed crystal.
anthracite coal: a hard, black coal.
aphanitic rock: rock with crystals too small to be seen by the naked eye.
aureole: the area next to an igneous intrusion where contact metamorphism of the original rock has occurred.
basic rock: igneous rock that contains 45 to 55% total silica.
bituminous coal: a medium-hard black coal.
borates: a group of minerals (e.g., borax) in which the borate structure (B03) is the main component.
botryoidal: resembling a bunch of grapes.
cabochon: a gem or mineral cut so that the stone has a smooth, domed upper surface lacking any facets.

carat: standard measure of weight for precious stones; one metric carat equals 0.2 g. The term also refers to a measure of purity of gold; pure gold is rated at 24 carat.
carbonaceous: composed largely of organic carbon (derived from plant and animal tissue).
clastic rock: rock made from pre-existing rocks and fragments.
clay: a sediment or deposit that is soft when wet and consists of very fine-grained materials.
cleavage: the tendency of some minerals to break along certain well-defined planes.
columnar crystal: crystal with an elongated prismatic habit.
conchoidal fracture: a type of fracture having long, curved surfaces, as in obsidian glass.
concretion: a concentration of mineral growth around a nucleus (matrix) into an orderly, rounded form in sedimentary rocks.
contact metamorphism: metamorphism caused by the intrusion of magmas at or near the molten rock.
crystal: a solid mineral mass having a regular geometric shape and bounded by smooth, flat surfaces.
dike: a wall-like body of igneous rock that cuts across the layers of surrounding rocks.
druse: a crust of fine crystals that fill a cavity or coat a matrix.
dull luster: luster with very little reflectiveness.
element: a basic substance from which all matter is made.
essential mineral: a mineral that must be present to classify a type of rock.
extrusive rock: igneous rock that is formed when magma erupts as lava onto the surface of the Earth.
feldspar: a group of abundant rock-forming silicate minerals, including orthoclase and microcline (potash feldspars).
ferruginous: containing iron.
foliated: having sheet-like layers with parallel orientation or segregation of different minerals.
fusion: the process of melting or dissolving by heat.
geode: a round cavity in a sedimentary rock lined with crystallized minerals, often in layers.
habit: characteristic appearance of a crystal in its predominant form.
hackly: a mineral fracture that has a rough surface with small protuberances.
hydrothermal vein: a vein formed by the crystallization of minerals from primarily hot water solutions of igneous origin.
igneous rock: a rock formed by the crystallization of a magma or lava.
inclusion: a solid, liquid or gas that is trapped in a mineral or rock when it is formed.
iridescence: the light phenomenon and play of colors.
lamellar: thin layers or scales composed of plates or flakes.
lamellar crystals: crystals developed in two dimensions (flattened).
lava: see magma.

luminescence: emission of light in response to a stimulus such as pressure, heat, rubbing, irradiation with ultraviolet light, etc.
luster: the reflective property of mineral surfaces.
magma: liquid or molten rock; it is called lava when it reaches the Earth's surface.
mammillary: aggregate of minerals formed with rounded shapes.
mantle: the layer of the Earth's major zonation that lies between the thin crust and the iron-nickel core.
mechanical weathering: a process in which rocks are broken down into smaller components with no change to their chemical composition.
mineral deposit: a concentration of one or more economically valuable minerals.
Mohs scale: a relative scale of the hardness of minerals, with arbitrary values from 1 to 10 (1 being the softest and 10 the hardest).
native element: an element that occurs in an uncombined form in the natural world.
orbicular structure: a rock structure consisting of large, rounded units containing alternating layers of minerals varying in type and color.
outcrop: exposed surface of an underlying rock body through the soil and debris.
oxidation: chemical reaction occurring when oxygen is added to a compound or process.
pegmatite: an igneous rock of extremely coarse grain size. Usually found as dikes within a larger plutonic or metamorphic rock mass.
phaneritic: an igneous rock with grains large enough to be distinguished by the unaided eye.
phosphorescence: the phenomenon of photoluminescence that continues even when excitation (usually by ultraviolet light) ceases.
pillow lava: rock with spherical forms, resembling pillows, taken on by some lavas when cooled rapidly under water.
placer: alluvial or glacial deposit where heavy minerals are recovered from the lowest portions of the sediment.
radical: a characteristic and essential constituent of a compound.
specific gravity: a measure of density using the ratio of the mass of a mineral to the mass of an equal volume of water.
streak: the color produced when a mineral is crushed into a fine powder; an important aid in identifying minerals.
subduction zone: plane along which the oceanic crust moves into the mantle as a result of plate motion.
tetrahedral crystal: a geometric shape having 4 triangle faces, 3 of which meet at each vertex.
ultrabasic rock (ultramafic rock): igneous rock that contains less than 45% silicates, a lower percentage than basic rock.
ultraviolet light: radiation with a wavelength slightly less than that of visible light.
vesicular: referring to a volcanic rock; having many cavities (vesicles) on the surface and inside.

Acknowledgments

I would like to thank several individuals, businesses and mineral museums for their kind help in putting this book together. Without their kind assistance, this book would not have been possible.

Individuals
Bob Hart (Princeton, BC)
Rose Wilson (Burnaby, BC)
Diane & Lum Zogas (Halfmoon Bay, BC)
Museums
Mineral Museum, Butte, Montana
Rice Northwest Museum, Hillsboro, Oregon
California State Mining and Mineral Museum, Mariposa, California
Pacific Museum of the Earth, UBC, Vancouver, British Columbia
Princeton & District Museum & Archives, Princeton, British Columbia
Businesses
House of Jade (Lillooet, BC)
Rock Shop (Lillooet, BC)
Lori Thomas, Manager, Princeton & District Chamber of Commerce (Princeton, BC)
Bruce MacLellan, Mountain Gems Ltd. (Burnaby, BC) www.mountaingems.com
Gloria Vickerage, The Fossil Shop (Drumheller, AB) www.thefossilshop.com
Bob Yorke-Hardy, Opal Resources Canada Inc. (Vernon, BC) www.opalscanada.com
Kaleidoscope Shop (Grand Forks, BC)

Bibliography

Beckwith, John A., 1972, *Gem Minerals of Idaho*, Caxton Press, Caldwell ID.
Chesterman, Charles W., 1979, *National Audubon Society Field Guide to North American Rocks and Minerals*, Alfred A. Knopf (Chanticleer Press), New York NY.
Farndon, John, and Steve Parker, 2009, *The Complete Illustrated Guide to Minerals, Rocks & Fossils of the World*, Anness Publishing Ltd,. London UK.
Hodges, Montana, 2010, *Rockhounding Alaska: A Guide to 75 of the State's Best Rockhounding Sites* (eBook), Globe Pequot Press, Guilford CT.
Hodges, Montana, and Robert Feldman, 1996, *Rockhounding Montana*, FalconGuides, GlobePequot Press, Guilford CT.
Hudson, Rick, 1997, *Field Guide to Gold, Gemstone & Mineral Sites of British Columbia, Volume I: Vancouver Island*, Orca Book Publishers, Victoria BC.
Hudson, Rick, 2006, *Field Guide to Gold, Gemstone & Mineral Sites of British Columbia (rev. ed.), Volume 2: Sites within a day's drive of Vancouver*, Harbour Publishing, Madeira Park BC.
Lynch, Bob, and Dan R. Lynch, 2010, *Arizona Rocks & Minerals: A Field Guide to the Grand Canyon State*, Adventure Publications, Cambridge MN.
Mitchell, James R., 2005, *Gem Trails of Northern California* (rev. ed.), Gem Guides Book Co., Upland CA.
Pellant, Chris, 1992, *Rocks and Minerals*, Eyewitness Handbook, Dorling Kindersley Ltd., London UK.
Romaine, Garret, 2010, *Rockhounding Idaho: A Guide to 99 of the State's Best Rockhounding Sites*, FalconGuides, Globe Pequot Press, Guilford CT.

Index